PRIMATE CHANGE

Yes; all the advantages of sitting still when he ought to move, and of leading a life of mere idle pleasure, and fancying himself extremely expert in finding excuses for it.

Jane Austen, *Emma* (1816)

For This Charming Siân,
(from a jumped up pantry boy)

An Hachette UK Company
www.hachette.co.uk

First published in Great Britain in 2018
by Cassell, an imprint of Octopus
Publishing Group Ltd
Carmelite House
50 Victoria Embankment
London EC4Y 0DZ
www.octopusbooks.co.uk

Distributed in the US by Hachette Book
Group, 1290 Avenue of the Americas
4th and 5th Floors, New York, NY 10104

Distributed in Canada by Canadian
Manda Group, 664 Annette St., Toronto,
Ontario, Canada M6S 2C8

ISBN Hardback 978-1-78840-022-0
ISBN Paperback 978-1-78840-078-7

A CIP catalogue record for this book is
available from the British Library.

Printed and bound in UK

10 9 8 7 6 5 4 3 2 1

Commissioning Editor: Romilly Morgan
Senior Editor: Pollyanna Poulter
Assistant Editor: Ellie Corbett
Copy Editor: Mary-Jane Wilkins
Proofreader: Helena Caldon
Index: MFE Editorial
Designer: Jack Storey
Typesetter: Jeremy Tilston
Illustrator: David Surman
Senior Production Manager: Peter Hunt

Primate Change

How the world we made is remaking *us*.

Vybarr Cregan-Reid

CONTENTS

WHAT BECOMES YOU?

Ill fares the land, to hastening ills a prey,
Where wealth accumulates, and men decay

Oliver Goldsmith, *The Deserted Village* (1770)

There is properly no history, only biography.

Emerson

If you think you're you, think again.

Your DNA, the code that instructs the assembly of the right amino acids into the right proteins at the right times in the right order and in the right places tens of trillions of times over, giving you a body, is not like a computer script. The code is not perfect, reliable or definitive. Instead, DNA sequences are more like the dialogue in a play. There may be a script, but the outcome depends on the environment in which those instructions are performed. Versions of *Romeo and Juliet* can vary in quality and tone, from the most ornate productions by the Royal Shakespeare Company to an elementary school performance.

DNA code is a little like this. With a few exceptions, we each have our own unique scripts, but the way those scripts perform in different eras or parts of the world shows considerable variety.

Our bodies are not just expressions of a code. We think of ourselves as the outcome of our unique genes, but our bodies also need an environment to shape them. For a body to function well it needs an appropriate habitat, and when there is friction or tension between body and habitat this can result in discomfort, pathology, disease and morbidity. A genetic code that is ill-matched to its environment spells trouble for the host. Organs that are highly adapted – the gills of a fish, for example – are useless in the wrong environment. DNA in the wrong environment will also have either to compensate in some way, or fail. And, now that we are at the point in our evolution as a species when we have already changed so much of the world around us, from its physical state to the way we interact with it, these changes are expressing themselves with chatty verbosity throughout our bodies, inside and out.

Because of the way DNA works, we are all different and we are all the same: identifiably human, but with an unending variety of shape and form. Every one of us is a complex genetic experiment, a random throw of the dice that hopefully suits the environment we meet when we first see the light of the sun. But because the environment plays such a key role in how our DNA is expressed, I guarantee that you are not the person that your DNA would have made 1,000, 20,000 or 100,000 years ago. Your height would be different, your weight, your face and your eyesight. And the chance that you would have an ingrown toenail, athlete's foot, asthma, seasonal or perennial rhinitis, tinnitus, acne, lower back pain, fatty liver disease, inflammatory bowel syndrome, short-sightedness, osteoporosis, a sleeping disorder, diabetes, depression, high or low blood pressure, rheumatoid arthritis, panic attacks, tuberculosis, Crohn's disease, fungal infections, malaria, attention deficit hyperactivity disorder, eczema, cavities and malocclusions, social anxiety, repetitive strain injuries, chronic obstructive pulmonary disease, and many cancers would be close to zero percent. This is what our environment has done, and is still doing, to us.

Still, modern life does have its benefits; in a busy metropolitan centre you are significantly less likely to be eaten by a dinosaur, so there's that.

OUR BODIES ARE OLDER THAN WE ARE

Humans are a little bit like plastic building blocks – indeed, any living thing is. Amino acids are the building blocks of life and these tiny organic compounds are pieced together to make proteins (of which there are millions of different kinds). Proteins go on to become cells, which in turn bind together to become tissue. Tissue makes up organs, and when they all function together, you have an organism. Whether it is a blade of grass or a beech tree, a starfish, a cephalopod, a Dilophosaurus or a human, the process is the same.

For this reason, the history of the human body is a long one, with some of our anatomical parts predating our species by hundreds of millions of years. But a step over two million years ago, the first identifiably human species began to walk the planet and since then several species of human have emerged, disappeared or converged. *Homo sapiens*, it seems, was the only species (among quite a few) sufficiently adapted to survive the harsh climatic changes of the Pleistocene and the Holocene. How they will fare in the Anthropocene is yet to be seen – you do have to worry about a species that names itself after the Latin word "wise". Indeed, some believe that the correct terminology for modern humans – those arguably more intelligent and technically capable than their forbears – should be *Homo sapiens sapiens*, so wise they use the name twice.

These anatomically modern humans (that most resemble us) first appeared in the middle Palaeolithic; the oldest known fossil discovered in Morocco in 2017 dates from 300,000 years ago. Although Neanderthals exhibited the kinds of symbolic behaviour that demonstrated their creative impulses, *Homo sapiens* were additionally thought to be capable of planning and abstract thinking (though evidence is mounting that this might also have been the case for Neanderthals).

The story of the emergence of these species and subspecies is one of evolution by natural selection, in which random mutation renders the organism more (or less) likely to breed and succeed in any given environment. But the processes and revolutions we are interested in are not evolutionary, they are cultural, and these cultural revolutions have led in turn to anatomical revolutions. During anatomical revolutions the body undergoes a number or series of changes that lead to an alteration of a feature or operation in response to one's surroundings or working patterns.

For the purposes of *Primate Change*, the key cultural revolutions or turning points in our development sometimes happened over thousands of years, and at others the changes might be measured in mere decades.

The first turning point is the very slow Agricultural or Neolithic Revolution that took place roughly 10,000 years ago in which humans shifted from being nomadic hunter-gatherers to become settlers and farmers. This Agricultural Revolution (*see* page 69) is not to be confused with others of the same name. It specifically refers to thousands of years of transition from hunter-gathering to the settled farming that is practised in cultures and countries throughout the world. But the demographic transition that took place was much greater than just the development of growing and production techniques. Hunter-gatherers morphed into a sedentary species that settled in villages and small towns, with all the associated environmental impacts attached to food production, such as irrigation and even deforestation. It is at this point, too, that we see the invention of something so basic that modern humans may wonder at it needing to be invented at all: storage.

Storage is only necessary when a society has reached a point at which it is acquiring and producing more than it needs – not something that previously troubled the *Homo* genus. First, *Homo* required the invention of something as simple as a jar. Then the jars needed to be housed on shelves – shelves go in cupboards, and so on. Fast-forward a few thousand years and this process has transformed into the fluorescent storage metropolises that skirt our towns and cities. These are places where we dump our things because we no longer have space for them, because the pleasure of acquisition is greater than that of possession.

The next turning point we might call the metropolitan or urban revolution, being the move from agrarian or village to city life. It was the second key point in our movement into the Anthropocene and the development of its body. From our perspective, this seems a long history, but this time of change is a good deal shorter than the Agricultural Revolution.

If the agricultural body is human version 2.0, then the metropolitan is version 2.1. Cities such as Damascus and Aleppo in Syria, Jericho on the West Bank, Athens and Argos in Greece or Plovdiv in Bulgaria have traces of settlements going back as far as 11,000 years, but they were not cities at that time. The ancient city of Ur in Mesopotamia is estimated to have had a population of about 65,000 people at its height – and this was about 4,000 years ago. Then, it would have been practically the only city that we would recognize as such. At the time, the world population is estimated to have been between 27 and 72 million, so the number of people living in cities compared to those living in rural communities would have been tiny. Fast-forward a few thousand years and the story changes substantially.

The tipping point comes in 1851.

In the middle of the 19th century, London was the capital of the world. The British Empire was about to reach its zenith, and the metropolis was the centre of culture and commerce for all of it, with the new northern cities being the engines driving manufacture and export. The current population of London is about 8.7 million. In 1851, it was just over 2 million. But that doesn't mean that the population density was all that different. The difference in numbers is mostly accounted for by the geographical spread of London throughout the 19th and 20th centuries. More people have spread themselves more widely.

During the 19th century, London's population grew quickly; from 2.2 million in 1841 it more than doubled in size by 1881 to 4.8 million. Population growth like this may seem normal to us, but what happened during those 40 years was part of a larger and more significant story that marked a change in our relationship with the landscape and the environment, disease, morbidity and everything.

In the 1841 census, the urban population was 48.3 percent of the country's population. But by the time of the next census, the scales had tipped. In 1851 this figure suddenly jumped to 54 percent. The significance of this shift is difficult to underestimate. There had been no previous time when a country's rural population was outnumbered by an urban or metropolitan one. And the trend continued; by 1891 urban populations had ballooned to represent more than 75 percent of the country's population. Throughout the 19th century urban populations increased by more than 20 percent during every decade.

The other spike in human history was, of course, the Industrial Revolution (*see* page 129) – the time when we transitioned from using hand production methods to devising mechanical ones. The industrial body is a new kind of human; version 3.0.

Inventions proliferated in the white heat of the foundries, with new technologies such as steam power, textile and paper manufacture, cement and concrete production, gas lighting, mining, the railway and other modes of transport conjured from the flames. The Industrial Revolution was the reason labourers moved from the country to the city during the 19th century and its impact was felt everywhere.

Innovation rocketed as quickly as the world's population soared. In 1750 estimates place the number of people at 700 million; today we have crested 7 billion and we are still accelerating, in much the same way as technology. It now takes just four and a half days for another million people to be added to the world's population and it is estimated that the total will reach 8 billion within the next decade. Estimates vary, but approximately 10 percent of all the humans who have ever existed are alive today.

At its most basic level, population grows because the birth rate outstrips the death rate, and advances in healthcare (particularly neonatal) also push up life expectancy. Today we can really see the rise of unnatural selection. With the refinement of healthcare techniques and technologies, a greater variety and abundance of food, as well as better education about lifestyles, infant (and parent) mortality rates have plummeted even in underdeveloped countries. In the 12 months between June 2016 and June 2017 the population in the UK rose at the fastest rate since 1947 – the year of the post-war baby boom. The reason for the recent rise? Worryingly, there is no particular reason beyond that of births outstripping deaths at unprecedented rates.

The impact of the Industrial Revolution on human bodies was significant. A broad range and variety of working practices existed before that period. I don't wish to romanticize them; many were hard, unpleasant, life-threatening occupations, but the range of body movements they required was extensive. Friedrich Engels, who studied the working classes of Manchester in the mid-19th century, suggested that the rural labourers of the preceding century had been lucky enough to have, "vegetated throughout a passably comfortable existence, leading a righteous and peaceful life in all piety and probity; and their material position was far better than that of their successors."^

During the shift from manual to machine labour, particularly during the first half of the 19th century, the range of body movement used by manual workers (or "hands", as they were called) diminished significantly. The work that the hands did, by hand, gradually became more limited and eventually became as refined as the carbs in a sugar sandwich.

There was much concern about the number of accidents and the terrible disabilities that were caused by modes of work in the 19th century. An avalanche of legislation was passed to protect the working classes (including young children) from widely-reported abuse by their employers. And the evidence cited was often the damaged bodies of the poor left unable to earn after work-related injury or disability, or as the result of chronic misuse – there were harrowing stories of workers having their skin stripped from their bodies by machinery, or being left blind or disfigured with amputated limbs.

What happened over the next hundred or so years was a further refinement of these new working practices, which was so complete as to be nearly unimaginable. Black smoke-chuffing chimneys sprang up in cities throughout the 19th century; they became a feature of the urban landscape everywhere, announcing these new ways of working.

We have something similar in the 21st-century urban landscape: the office block. These new buildings announce version 4.0 of the human body. They declare an even further refinement of working practices. These are undoubtedly healthier environments than the factories that preceded them, but their plate-glass windows, strip lighting and stain-proof carpets hide perhaps the most significant contributor to our Anthropocene bodies.

This second wave of the Industrial Revolution is not quite as unhealthy as the first, but it is persistently toxic. What I am doing as I write this is what more than half the population of the world is probably currently doing: sitting down. The majority of work now undertaken on the planet is performed in this pose, with the body bent at two right angles while sitting on a chair, hunched slightly forward, with rounded shoulders and the nuchal ligament at the back of the neck engaged to stop the skull crashing onto the keyboard.

Despite what we might think, chairs have not been common for all that long a period. While they are extremely old, chairs were historically rare. There are monuments in the ancient Babylonian city of Nineveh that feature ornate chairs. There are images of them on Greek vases and carved steles, as well as representations of them in early Chinese and Japanese culture. For millennia they have been used as a symbol of authority (the highest academic attainment in my profession is referred to as "a chair") and perhaps because the chair had become such an important status symbol, for a long time it could not be widely adopted.

In homes, benches and stools were widely used for many hundreds of years, but it was not until the Early Modern period that chairs became a common feature of a family's home – hard, wooden, upright chairs were mostly to be found.

For all its tough ways of living, working and being, the Industrial Revolution was also a period during which people craved comfort and leisure. During the early 18th century, there was a shift in the fashion of manners in the French court, from upright formality to a much more leisurely mode of social interaction in a seated, semi-reclining posture. This fashion went hand-in-hand (or bum on seat) with the introduction of the fully upholstered chair, one in which it was comfortable to sit for hours at a stretch.

Add to this the invention of cinema, TV and video games and you have a perfect storm of sedentary work followed by predominantly sedentary leisure, too. The imbalance is so extreme that the fingers of a sedentary worker roam up to several kilometres a day as they dance briskly across phone screens and computer keyboards; while some studies suggest that their feet cover 1km (0.6 of a mile) *over an entire month.*^

None of us can sit down for eight hours a day experiencing all the terrible effects that inactivity has on our bodies, and then expect a 30-minute yoga or spinning class to magically undo the marble-like stiffness we are instilling into our bodies.

The muscles in the back of sedentary workers gradually become enfeebled from lack of use; the little work they do is reduced to occasionally carrying the 30–50 kilograms (66–110lb) weight of the torso. But when sitting they don't even do that. When a body has weakened musculature, it is easier for dysfunction to creep into the joints, which through movement, force and reduced resilience can alter the mechanics of the spine.

In *The Analysis of Beauty* (*see* below), 18th-century artist William Hogarth wrote about a "waving and serpentine" S-shape as "the line of beauty", which "leads the eye a wanton kind of chace".^ At the core of this master's inquiry lay some simple principles of balance, symmetry, variety, simplicity, regularity and intricacy, which for him the S-shape perfectly embodied. The human spine possesses a similar line of beauty because it is also highly functional. The shape maintains a good centre of gravity during bipedal movement, as well as functioning as a kind of spring that can absorb shock in both running and walking. But by the age of 40, most of us who have done sedentary work are unable to stand up correctly because our perfect S-shape is starting to look a little more like a Z. (An S-shape absorbs vertical shock much more efficiently than a Z does.)

The wave-like, serpentine, perhaps spinal, lines of beauty are scattered throughout "Plate I" from Hogarth's, *Analysis of Beauty*.

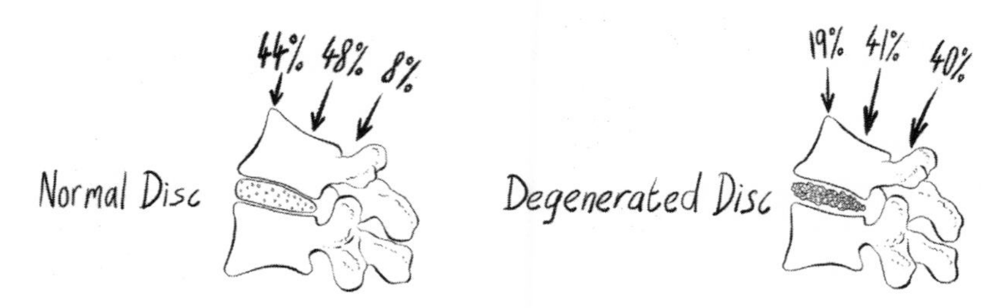

An example of how spinal loading can change with age, based on Mike Adams and Patricia Dolan's research in "Spine Biomechanics", *Journal of Biomechanics*, 2005.

When the S-shape becomes a Z in an ageing sedentary worker, the discs at the base of the spine are strained because of uneven pressure – and because they are less elastic than they were during the preceding decades. These changes to the stress-shielding capability of the intervertebral discs can multiply pressure toward the posterior edges of the spine by up to 500 percent between the ages of 20 and 40.^To make matters worse, the newly shortened muscles might also gradually pull the pelvis out of position, so that even when standing, it can be difficult to find a stance that is not pathological in some way because the mechanics of the spine have been altered.

Many sports and physical therapists suspect that when shortened muscles pull the lumbar spine out of line a group of muscles called the hip flexors are the root cause. This process of muscle shortening takes a considerable length of time – it is geologically slow, but then we give it a lot of time. We sit for an estimated 70–100 hours a week – that's four to six years of every decade; longer than we spend sleeping, or doing anything else, in fact. A session of Pilates once a week which spends only a few minutes on stretching hip flexors is unlikely to do much to alleviate this process, the momentum of which builds with the threat of a gathering iceberg drifting in northerly latitudes. Our bodies are destined to hit this iceberg some time in early middle age and most of us are powerless to stop it.

The first time I experienced back pain was when I first started using computers in the early 1990s, I was simply required to put the right numbers in the right boxes, execute some commands and move on to the next item (the very essence of modernized factory labour). I had an office chair (a new experience for me), my desk was near a window out of which I had a pleasant view. As offices went, it was a good one. This was a simpler time: mice had balls, men's shorts had just two pockets, and in-trays strained under the weight of memos and internal post because no scoundrel had yet invented email. For some reason, on the wall above my desk was an A3 diagram. This was a line drawing of what looked like a crash-test dummy; instead of joints it had large, circular hinges, with lines indicating range of movement. Its back was gun-barrel straight, and its feet rested in a parallel and unfidgety pose on the floor.

It was a version of me (and those temps who had preceded me), seated at a desk, locked in a rigid and seemingly perfect posture – a cyborg, an ideal synthesis of human and machine in which biological and mechanical technologies met at laser micrometer-perfect 90-degree angles. The worker in the diagram was a model employee, never stopping, chatting or slouching. The worker never shifted in their seat or gossiped over their cubicle. The eyes of the worker were always trained on the monitor, and like an ideal partner, the worker's height matched perfectly that of the computer. This rigid stillness protected the worker from harm because of "good" posture.

The fiction of the diagram was known to me then. No one works like that, where everything they do is ergonomically optimized. Those who sit in a fixed posture cannot avoid RSI, eye-strain, sciatica or any of the hundreds of other pathologies known to be associated with sedentary living and working.

This lifestyle was the creator of my back problems. It wasn't that my body was somehow too weak for modern life, but that the human body was never meant to physically experience modern life in this way. Today, remaining static and rigid in any kind of posture is one of the few known causes of back pain.

This is not my story, though.

I am one of the billions whose body has been changed by modern living. And we have much to be grateful for: central heating, beds, longevity – our standard of living is better than that of a medieval monarch. Many other items appear in the list of modern life's benefits: analgesic dentistry, transport and billions of prescription drugs – of course we should be thankful for these, too – but our gratitude should be tempered by remembering that the majority of the problems they are addressing only exist because they have been created by the way we live now.

The chances are that most of the people reading this will not die of natural causes, but of a mismatch disease; this is not just the result of being born with the right (or wrong) DNA. Mismatch diseases are thought to be brought about by the tension between our bodies and the newness of the environment that those bodies are required to inhabit. And they are mostly diseases that are familiar and seem natural to us. Type 2 diabetes, for example, has been around since the dawn of humanity, but in the environment and with the diet of the Palaeolithic hominin there was little opportunity for the genetic disposition to manifest itself. There was little processed food and sugar to exacerbate the condition then. Fast-forward two million years, and the same genetic predisposition exists in a more toxic environment encouraging overconsumption, in which it is cheaper to buy a bag of jam doughnuts than an avocado.

While early humans probably consumed about ten tablespoons of sugar per year, in the modern Western diet that is more likely to be the amount consumed daily.

If we try to plot a graph or draw a pie chart to show how astonishingly swiftly all these changes are happening, how quickly our environment is changing, how recently our bodies have become (as Hamlet said) "out of joint" with their times, the proportions present us with some difficulty. In the pie chart opposite each of these revolutions is included, but the only visible one (when compared to the entire lifespan of our species) is the 10,000 years during which we have been farming. The overwhelmingly vast majority of our history has been spent hunter-gathering. The time during which cities have risen, or since the Industrial Revolution, or the time in which we have been doing sedentary work; we think of them in human terms and they seem normal, even traditional and longstanding to us – they have been around for generations. But they are all such a small part of our history that we cannot see them. They do not even exist as a single pixel on a computer screen.

To put it another way, if the timeline for human-like species were condensed into a nine-to-five working day you would need to wait until 4.58pm for the Agricultural Revolution. Even the smallest cities weren't built until well into 4.59pm. The Industrial Revolution? You would have to have been keen-eyed to notice it. It *began* at 4:59 and 58 seconds. Nearly all the technology we know and interact with would have come and gone in about the time it takes to sneeze.

As a result, our bodies are in shock from these changes. Modern living is as bracing to the human body as jumping through a hole in ice. Our bodies are defending and deforming themselves in response. Our limbic systems pump us full of nervous tension because they find themselves in urban environments they do not recognize. These environments seem normal to us on a conscious level, but on an unconscious one they lack traditional sources of food or water, and so, as far as our primitive brains are concerned, they

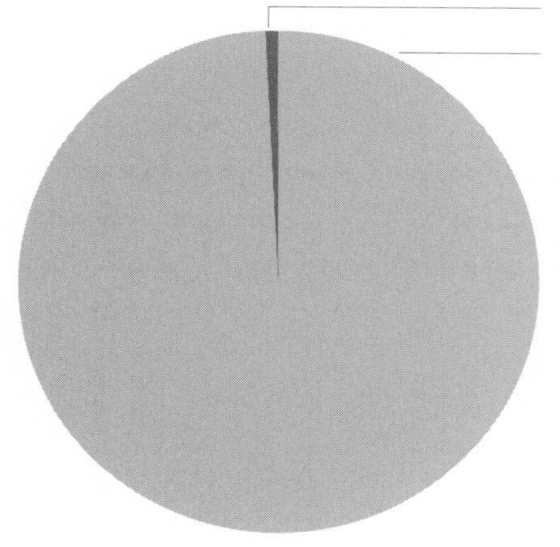

In this pie chart both the Metropolitan Revolution and the Industrial Revolution were included in the calculation, but their size relative to the lifespan of our species is too small to see in print.

believe we have chosen to settle in a spartan desert.

As the fountain of innovation continues to flow in modern life, through GM crops, the acceleration of virtual-reality, augmented-reality and artificial intelligence, and as more jobs are snatched by robots, new pathologies will emerge as our bodies crave to be outdoors doing things they recognize as simulations of hunting and gathering (the activities our bodies evolved to carry out).

These desires are independent of our rational thought processes – and these are the activities for which the body receives the richest rewards. The sheer number of them is breathtaking. Perhaps there are ways in which we might meet these needs. In an odd way, they are inscribed with the definition of tattoos all over the Anthropocene body. From our toes up to our knees and hips, through our joints, pelvis and chest cavity, around our shoulders and along our arms to the tips of our fingers, along our curving Cyrillic spines, across our misshapen faces, to the tops of our drooping heads, changes are written all over our bodies. We just need a cipher to be able to read them, and then we can start learning how to reclaim and restore our modern, sagging husks.

The examples included here have been chosen in some cases based on my own body and its relative commonality, being that of an unremarkable Western male, neurotypical, able-bodied but beginning to be restricted by middle age – and with pathologies shared by millions, if not billions. A complete analysis of the modern body would require a library of books on everything from breast-feeding, the contraceptive pill, cosmetic surgery and antibiotics to every kind of cancer, neurodegenerative disease and allergy, so the focus will be mostly on how particular body parts shift in and out of focus at key points in our history, or on what those key moments have *done* to parts of our bodies. Diet also seems to be a subject the Anthropocene human is obsessed with, and it sits mostly in the background of many of the chapters here for the simple reason that the shelves of bookshops are already bowed under the weight of the tomes published on the subject, with everything from the "red wine" diet to urban "paleo" recipe books.

And so, with some sense of scale, let's head back in time to make a little more sense of who it is we are becoming.

THE ANTHROPOCENE BODY

Our appearance and the way we move, rest, sleep, think, eat, gather and communicate have all changed dramatically since *Homo sapiens* first walked the planet probably well over 300,000 years ago. We have not evolved all that much since then, but as we have been extremely busy farming, irrigating, planning, building, mining, drilling, testing and dumping waste in those surroundings, those very surroundings have been changing us. And we are arriving at a moment in our history that is about to be named for the tremendous impact our activities as a species have had on our environment.

About a year ago I was talking to a class of English Literature students about Dickens and urban life and I asked them a relatively simple question – one they may have been able to answer in secondary school, but to which they had since forgotten the answer. It was also a loaded question: "Which geological epoch are we living in?"

Geological epochs were defined in the 19th century, although at the time those who named them had no idea how far back they stretched. They thought they were dealing in a currency of thousands, perhaps millions, of years. With the discovery of radiometry in the early 20th century the geological history of the Earth has since been traced back 4.5 billion years.

Thanks to the endeavours of history's geologists there is at least one right answer to my question, but there is another one, too. The first and more traditional answer my students could have given me was that since the end of the last Ice Age about 11,700 years ago we have been in the Holocene epoch. It is characterized as a relatively stable and warm phase of Earth history after a glacial period that lasted about 100,000 years. One of the stranger things about the Holocene is that it is a relatively short span of time – the preceding epoch, the Pleistocene (during which humans evolved) was a staggering 2.5 million years in length.

The last Ice Age was hard on the human body; there were at least 20 cold cycles of freezing and thawing, and on average global temperatures were about five degrees cooler than today. Consequently, the planet was much drier as there was much less atmospheric water because so much was frozen in colossal ice sheets. This was a hard environment for *Homo*, and if those deep cold snaps had not happened there could still be several different species of humans on the planet today.

The second option the students could have chosen was that we are living in the Anthropocene (from the Greek words *anthropos*, meaning "human" and *kainos*, meaning "recent" or "new"). The term was coined just a few years ago by the Nobel prize-winning, atmospheric chemist Paul Jozef Crutzen (although as early as 1873 a similar term was used by the Italian geologist Antonio Stoppani, who called it the anthropozoic era). Although the term Anthropocene is not yet widely known, it soon will be. Within a year or so of this book being published, it will be officially adopted. During the following decade, kids will return from school telling their parents about it. In ten years, the man sitting next to you on the bus will know what it is. In a hundred years' time, people will still be thinking, writing and talking about it; and in five hundred years... but perhaps we should not get ahead of ourselves.

The name is teetering on the brink of official designation. The people who debate and decide on the timing and naming of such things are the members of the International Union of Geological Sciences (IUGS).

The group was founded in 1961 with the goal of establishing international cooperation in the field of geology. In 2009, a working group was assembled which was asked to gather evidence of the Anthropocene. The members of the IUGS are understandably reluctant to tinker with their hard-won geological timescales. The chronostratigraphic table, with its divisions, subdivisions and sub-subdivisions of supereons, eons, eras, periods, epochs and ages is as beautiful, simple, concise and encyclopaedic as the periodic table is in chemistry. There is not only a story behind every name on the table, but a complete and unrecognizable version of our world, too. Just as the editing of the periodic table is not taken lightly, the idea that we are living in a new period is considered hugely challenging because the evidence used for establishing a new geological epoch cannot just be local, but must be found globally.

The working group has proposed that there is overwhelming evidence that the planet, its atmosphere, its oceans and wildlife have been permanently changed by humans. The changes we see today are as substantial as those made by the last Ice Age. Much of the evidence is based on changes that we are unlikely to brush up against in our day-to-day lives. Looking for that global evidence in this new epoch, there is a sudden spike in mineral novelty on the planet (human ingenuity has devised many new compounds). Radioactive isotopes from numerous nuclear tests have also been found across the planet. Dangerously high levels of phosphate and nitrogen in soils (from artificial fertilizers) also provide evidence. Then there are the problems that are all too recognizable to the Anthropocene human: plastic pollution, the globally pervasive spread of concrete particles – even chicken bones can be included as evidence because the remains of the billions and billions of chickens that have been produced for human consumption are fast becoming a permanent part of the fossil record. The modern chicken that is industrially reared for human consumption is also larger and meatier than it was just a few decades ago. Thanks to our endeavours, we now boast Anthropocene chickens as well as humans.

This sounds beyond the realms of reason; however, it stands as a stark example of our own bodily transformation. The size, shape and bodies of animals have changed significantly as a result of human intervention. These changes could seem as innocent and insignificant as those that influence the breeding of dogs, but they tell us how easy it is for complex organisms to change through a little reproductive management and animal husbandry. The story of the changes in humans follows the same narrative thread, although it is a little more complex.

The bodies of humans evolved during the Pleistocene between two and three million years ago and have moved through several stages to become what they are today. All of us carry a hotchpotch of genes that cannot be said to be in any way "human" because so many of them are inherited from older life forms and species.

Inside a human cell is a nucleus; inside the nucleus are (with some exceptions) 46 chromosomes – 23 from each parent. On each chromosome there are thousands of genes, and if these were removed from most of the cells in your body (red blood cells and the lenses in your eye are examples that contain no DNA), and uncoiled, their length would take you far, far beyond Pluto and out of the solar system altogether. We carry *a lot* of DNA: genes have to make many different kinds of proteins that make parts that make parts that make parts to eventually form body parts. In a system so large and complex, genes are often selected for over thousands of years, but cultural changes, as we shall see, can be introduced in a generation.

One of the key moments in our evolutionary history was when, in different parts of the world, we stopped collecting food and began growing it instead. Rice became a crop in China around 11,000; chickpeas, lentils and others (called founder crops) around 9500BCE in the Middle East. In Mesopotamia (modern-day Iraq) pigs were domesticated as early as 13,000BCE, sheep between 2,000 and 4,000 years later and cattle much later, around 8000BCE. It seems odd to call a period of about 4,000 years a revolution, but that is what it was – an Agricultural Revolution.

As humans significantly changed their relationship with their immediate environment by farming instead of hunting and growing instead of gathering, their bodies began to change, too. Their new diet didn't just change the shape of their stomachs, but also their faces. The number of teeth they had (and we still have) became surplus to requirements. With their softer and mealier diet their jaws failed to mature and expand properly, so they developed malocclusions, or misalignments. Their teeth didn't fit in their head any more. The shift to a carbohydrate-heavy diet also brought with it more tooth decay. Our genes respond to some of these changes where and when they can, but they are doing so inconsistently and very slowly; in other cases, evolution plays no role whatsoever. Evolution does not take into account health and wellbeing – or pain or morbidity if these strike beyond the age at which humans normally reproduce. And 10,000 years (as shown on the illustration on page 17) is a tiny span of time when compared to the history of the species.

But in that short window of time we have been busily changing the world; altering its lithosphere, messing with its species, polluting its oceans and drilling holes in its strata – Kola, a 12-km (7.5-mile) deep borehole was built out of curiosity by the Russians. We have also been experimenting with all the responsibility of a toddler brandishing its parents' loaded gun. The Anthropocene human is one whose body has changed – not as a result of evolution, but in response to the environment we have created. With new scientific discoveries, experiments in living, changes in modes of working, alterations to our social landscapes and countless other transformations, improvements and innovations, the world we've created has quietly been changing us, too.

Back to the seminar group; although I was discussing Dickens and epochs with them, what I was mostly thinking about was the scalding pain in my back. I've suffered with it intermittently since my twenties, but there are times when I have it for such long periods that I believe it has at last become permanently, painfully, mood-alteringly chronic. As I walked and talked during the class, I occasionally had to stop and crouch in a full squat to relieve the pain.

I creaked my way around the classroom as we started to talk about the bodies of some of Dickens's characters from his last novel, *Our Mutual Friend*.^ It is a voluminous book, a banquet of social ills in which all the metropolis appears, from the highest echelons of the new-monied classes with the society-loving Veneerings, to the precariat existing on the edges of civilization. There is Mr Venus, sentimental and creepy in equal measure. He is an articulator of limbs who deals in body parts, both real and prosthetic. Silas Wegg, a street trader of ballads, is desperately trying to earn enough to buy back his amputated leg from Mr Venus's eerie emporium. And there is Jenny Wren, a twisted and stunted doll's dressmaker who is the child-parent of an alcoholic father. At a number of points through the novel she tells us, "I can't very well do it myself, because my back's so bad and my legs are queer." It is a novel of a society as mutilated and mangled by its conditions as Jenny Wren's spine.

However, before the 19th century, bad backs do not make much of an appearance in literary history. Famously, there is the case of Shakespeare's Richard III, who we have seen hobbling and hunching his way about the stage for four centuries. In Thomas More's *History of King Richard III* (approximately 1513–18), he is described as: "little of stature, ill-featured of limbs" and "crook-backed".^ But these descriptions are tied up with early-modern ideas about biological determinism – the notion that one's personality is somehow inscribed in one's physical appearance – so his ethics seem as crooked and twisted as his spine. By Dickens's time, this had changed. While the Victorians had their own versions of biological determinism (just as we do today), the explosion in both the variety and instances of disability is explained by the revolutionary change in working practices. As the mode of work changed, so did the work-related disabilities associated with it.

There is a tendency today to think of our work as easier than work done in the past. I certainly feel that my job cannot be considered hard work. I get stressed, as anyone does, but my grandparents ran a farm. That was hard work. I stand at lecterns in centrally-heated and air-conditioned spaces; I stroll around seminar rooms and write some emails; yet there are times when my back pain is crippling. And I am far from unusual in this.

Back pain, specifically lower back pain, is now the single leading cause of disability throughout the world. It is the most common reason for missed days of work. It is the second most common reason for visiting a doctor. Half of all adults in the US have had symptoms of back pain in any given year, with an estimated 80 percent of adults expected to suffer from it at some point in their lives.^

The spiralling cost of healthcare connected to these common modern pathologies, such as back pain, type 2 diabetes and chronic obstructive pulmonary disease, are now so acute that with the addition of a little more time to gather momentum, they will be enough to bankrupt the health services of at least a few countries. While it is true that we are living longer in the West, as longevity inches upward morbidity is skyrocketing.

If this all sounds rather grim and hopeless, help is coming. While there is no "solution" to the Anthropocene, there are sackfuls of them for the Anthropocene body. If much of our morbidity derives from our lifestyle, then in many cases it is a matter of making simple changes that will have a *big* impact on the way we live. The solutions usually focus on giving our bodies a little more of what they were expecting to find in the environment they were born into, rather than the weird concrete land they encountered. These solutions are not to eat raw meat or drink from a river, but more about finding ways of making the benefits of modern life work better for us. At the end of each chapter of the book, there is a short section entitled "Winding Back" with some nudges and a little advice, grounded in research, to help relieve some of the tension that exists between our bodies and their environment. Because there are incrementally more changes in the environment, these "Winding Back" sections grow longer as we move through the different versions of the human body. But, for now, let's find out a little more about how we got into this mess by looking at the kind of body that many of us might wish to wind back to inhabit.

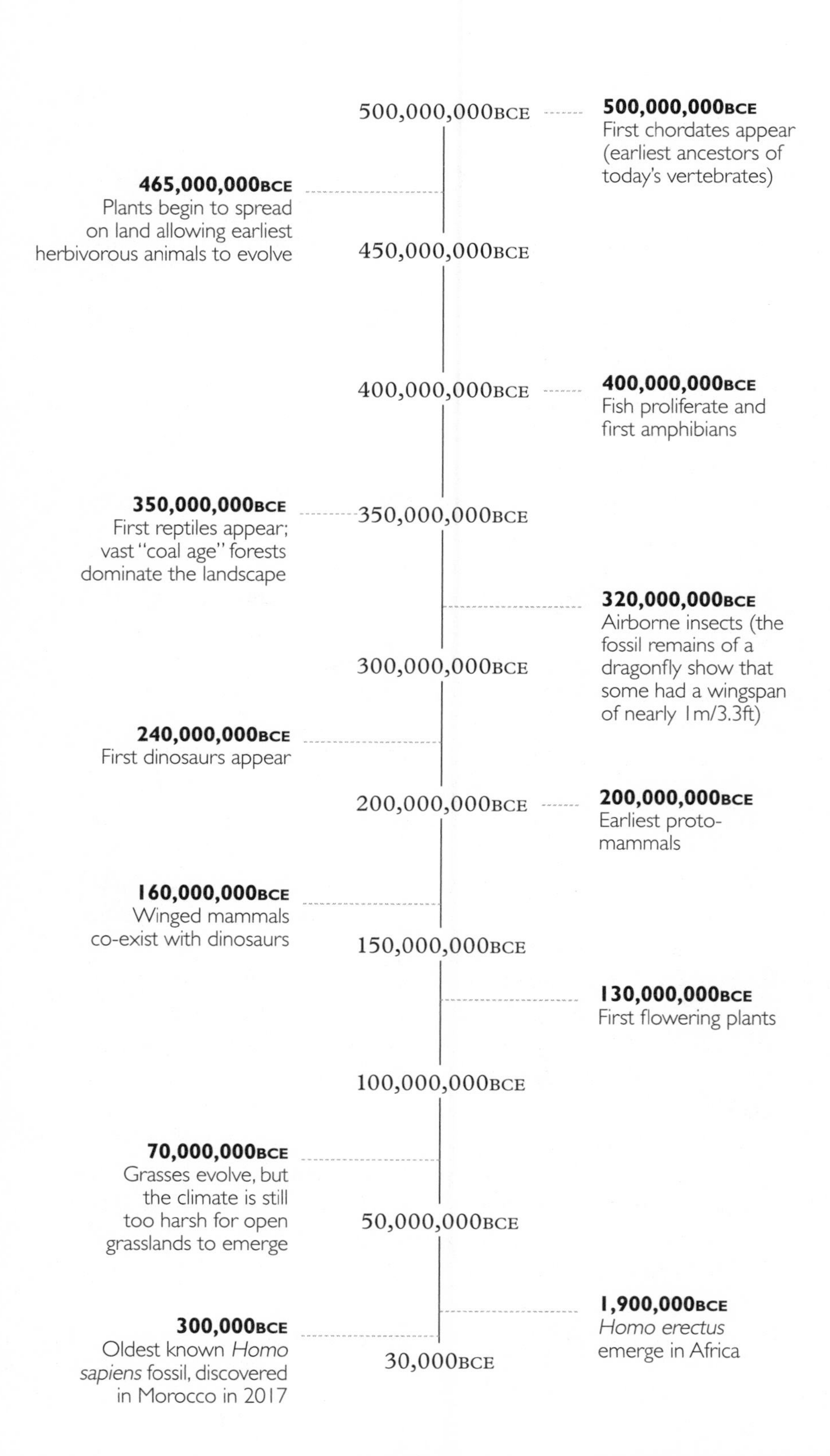
500,000,000BCE
500,000,000BCE
First chordates appear (earliest ancestors of today's vertebrates)
465,000,000BCE
Plants begin to spread on land allowing earliest herbivorous animals to evolve
450,000,000BCE
400,000,000BCE
400,000,000BCE
Fish proliferate and first amphibians
350,000,000BCE
First reptiles appear; vast "coal age" forests dominate the landscape
350,000,000BCE
320,000,000BCE
Airborne insects (the fossil remains of a dragonfly show that some had a wingspan of nearly 1m/3.3ft)
300,000,000BCE
240,000,000BCE
First dinosaurs appear
200,000,000BCE
200,000,000BCE
Earliest proto-mammals
160,000,000BCE
Winged mammals co-exist with dinosaurs
150,000,000BCE
130,000,000BCE
First flowering plants
100,000,000BCE
70,000,000BCE
Grasses evolve, but the climate is still too harsh for open grasslands to emerge
50,000,000BCE
1,900,000BCE
Homo erectus emerge in Africa
300,000BCE
Oldest known Homo sapiens fossil, discovered in Morocco in 2017
30,000BCE

Part I

500,000,000–30,000BCE

PRIMATES CHANGE: MOVEMENT, MECHANICS & MIGRATIONS

> … that man with all his noble qualities, with sympathy which feels for the most debased, with benevolence which extends not only to other men but to the humblest living creature, with his god-like intellect which has penetrated into the movements and constitution of the solar system – with all these exalted powers – Man still bears in his bodily frame the indelible stamp of his lowly origin.
>
> Darwin, *The Descent of Man* (1871)

I used to love nostalgia.

Looking back to previous ways of life is a perilous occupation. It is all too easy to filter out the hardship and discomfort of the past, and even as M R James observed, its ugly brutality. "A View from a Hill"[1] is a typically creepy story he wrote in which an ageing antiquarian, while visiting an old friend, finds that the blurry binoculars he has borrowed are a window to the past. At first, he sees green fields, meadows and maypoles, but a startling part of the view from the hill of the title is a gibbet with a body hanging from it. Now part of an important genre called folk horror, this James story is as much a warning to the nostalgic as it is to the curious; a little nostalgia can be a dangerous thing.

As we get up onto two feet we must tread carefully when looking at the history of the human body. The tenor of the argument here is not to suggest that we can go back in time – if we could, who would want to? But there is

1 From a 1925 collection *A Warning to the Curious and Other Ghost Stories*.

a great deal that we can learn about ourselves by looking at where we came from, and it is often the case that remedying some serious ills can be a matter as simple as going for a brisk walk or playing out in the sunshine. The DNA of the Palaeolithic human expected this of us, and it still does. This section focuses on the body as it evolved, imperfect certainly, but nonetheless one that was adapted to revel in certain behaviours, and one that was rewarded for fulfilling even the most mundane tasks, such as walking.

ANIMALIA CHORDATA

Where did our bodies come from? Why were some random variations in the history of our species so favourable to our survival? The answer to these questions are not to be found only in the shape, form and function of our bodies, but in the environments they encountered when the Earth was a different place. Fins may generate tremendous locomotive power, but not in the air or on land; instead, a body *and* its environment must meet for variation to be favourable. We need to glance at the deep history of the human body, looking at how primates changed during the Pleistocene, and how some of the body parts that we *Homo sapiens* inherited might not be suited to our modern world. To understand what is happening to our bodies today, we need to understand where they came from, and what difficulties their adaptation was addressing. What is the human body for? What tasks does it need to fulfil in its environment? How might those tasks have changed?

The human body was never perfect, not during the Industrial Revolution, nor during feudalism when peasants worked the land, nor when cities first rose from the sands of what would become Mesopotamia and Egypt, and not during the Agricultural Revolution about 10,000 years ago, or even before then. But without being nostalgic, we know that at the very least the human body was strong enough for us to shuffle to the top of the food chain after what some palaeontologists have called the Cognitive Revolution. How did we get there? How did our early bodies adapt to make best use of the environment?

It was a journey – such a long journey – and we made it all on foot; everything held together by a firm, strong column, the fulcrum of the multidimensional gait cycle: the spine.

The name by which our species is classified runs thus: *Animalia Chordata Mammalia Primates Haplorhini Simiiformes Hominidae Homo sapiens.*^ The first refers to the fact that we are multicellular with the ability to move. The second is to do with our spine (and the cartilaginous notochord that runs along our backs in the womb).

These bones in our backs have a staggeringly deep and broad history. According to palaeontologist Stephen Jay Gould, they have formed the core structure of trillions of animals for hundreds of millions of years, an idea he explored in *Wonderful Life: The Burgess Shale and the Nature of History*.^

The Burgess Shale is a particularly fruitful deposit of fossils in the Canadian Rockies of British Columbia, Canada; they date to about 500 million years ago, just after the Cambrian explosion – a period which saw a relatively sudden proliferation and divergence of life forms on the planet. The phenomena was observed as long ago as the 19th century; geologist William Buckland noted (and he was not alone in doing so) that the sudden appearance of organisms such as trilobites, for example, left Charles Darwin with out-turned pockets, gesturing to us with empty hands, when, in 1859, it came to explaining this phenomena in *On the Origin of Species:* "To the question why we do not find rich fossiliferous deposits belonging to these assumed earliest periods prior to the Cambrian system, I can give no satisfactory answer."^

Gould's book argued that in the Burgess Shale there was a greater diversity of body blueprints (phyla) than can be found today. Morphologically, it was a distinctly more experimental period for life on Earth. Gould is, of course, not the first palaeontologist to work on the Burgess Shale, but he was terrific at pulling the significance of this fossil record into focus with his thoughts on the meagre and diminutive Pikaia worm. The fossils themselves (there are many) show an animal, a little flattened, like an eel, but only about 5cm (2in) in length. But unlike, say, a slow worm, the fossil is not smooth, instead it has segmented blocks of parallel muscles that look not unlike a spine. Its significance is such that a team headed by Professor Simon Conway Morris of Cambridge was able to establish, by looking at more than a hundred fossils of the worm, that it possessed a notochord (that flexible cartilage found in the embryos of all vertebrates).^

The conclusion is that the Pikaia worm is the common ancestor of all vertebrates on the planet.

About half a billion years ago during the Cambrian explosion, early chordates (organisms with a hollow dorsal nerve chord) began to appear. The years in their millions marched onward and during the Devonian period (the age of the fishes about 419–359 million years ago) these organisms had skulls, jaws, fins, tails and gills. As the Devonian gave way to the Carboniferous (359–298 million years ago), fossils show, through the developing robustness of their bones and muscle attachments, that some life forms were becoming more committed to venturing onto land. Even today, some fish use their limbs in the way a mammal does. The Sargassum frogfish, in order to evade a predator, deploys digit-like appendages to clasp surrounding vegetation. How and why those limbs later developed five digits as the ideal across species, scientists still seem unsure.

About 70 million years later, limbs began to appear more frequently below, rather than on the side of the body. This led to a burgeoning predominance of reptiles (Mesozoic, the name for the geological era, literally means "mid-life", between fish and mammals). This limb development is still evident in some fish such as mudskippers or the Pacific leaping blenny, which have feet-like fins.

Farther on, from about 200–66 million years ago dinosaurs flourished. Early mammals (the "mammalia" of our classification, being warm-blooded, featherless, with teeth, and so on) coincided with them, but they did not begin to thrive until much later, when the Mesozoic era became the Cenozoic (or the age of mammals, as the word means "new life"). In this era, about 55 million years ago, dry-nosed primates (the "Primates Haplorhini" of our classification) appeared and spread throughout the world. These monkeys and apes (the "simiiformes" of our classification) had collarbones, fingers, thumbs, toes and forward-facing eyes (unlike dolphins and whales) and flourished in and around the trees of Africa.

About 20 million years ago, in an unrecognizably different climate, Africa was basically a super-forest, an enormous planetary lung exhaling tonnes of oxygen into the Earth's atmosphere. In such an environment it is hard to imagine an evolutionary advantage to emergent bipedalism, but there was still migration. Early apes such as *Proconsul africanus* (discovered in 22-million-year-old deposits in Kenya) and *Dryopithecus* both made it across the land bridge between Africa and Eurasia and had spread across Europe and much of Asia 10 million years ago. These were the early great apes (or *Hominidae* – no tail, S-shaped spinal column) that eventually led to *Homo*.

When we are dealing in millions of years, we are not just interested in the life and death of organisms, but also the persistently changing face and behaviour of the planet itself. For the previous 200 million years, India had been drifting north through the Indian Ocean and began colliding with Asia about 50 million years ago. As the collision persisted about 10 million years ago, the formation of the Himalayas was so significant that it had an impact on the world's climate. Their height made India a much wetter place, subject to regular monsoons, drying the surrounding air so that as the winds drifted to meet the super-forest of Africa, they were no longer bringing rain with them. About 5–10 million years ago, the forests began to dry out, dying back to leave wide open grasslands exposed to the sun.

Bipedalism may not have offered any advantage to an adapting species of the forest, but that was not the case for grasslands. The sudden emergence of these two different habitats brought with them a species split between those adapted for the trees and those for the grass – and neither environment is particularly good for the preservation of fossil remains.

Our hands and feet, unique among species, were not a huge evolutionary novelty. Although the correlation is complex, a set of genes called the Hox sequence, which contain code for how different structures develop, is found in all animals, from giraffes to jellyfish, blue whales to fruit flies. Much animal diversity is founded on the simple principle that bodies are constructed from repeating units similar to plastic building blocks, with genetic programmes for building particular structures into certain compositions. In the animal kingdom, Hox genes are activated in different ways to create fins, antennae, peculiarly long necks, heads, tails, and of course, our feet.

Our own feet are basically tetra-paddles, adapted (as fins are) for locomotion and negotiation of solid ground as opposed to water, so it's not a surprise to find that they are constructed using the same programs and cells that made fins for many millions of years. Palaeontologists are unsure why fins became limbs in the earliest land animals, but the most likely environment to favour this adaptation is a shallow swamp where food sources and opportunities broadened for those with sufficiently strong limbs to venture onto land.

We might celebrate Columbus or Magellan, but transitional tetrapods were surely the true pioneering explorers.

Our success as a species cannot be accounted for by any single body part, but our feet play a key role. And we are still in the process of trying to understand their importance both to modern and early humans.

FOOTPRINTS

The oldest beta-human fossils tend to be cranial; the paucity of data for the other end of the body exists for a few reasons. As a species, we probably only began burying our dead about 100,000 years ago. Ancient remains, then, have to have met with several fortunate accidents to be preserved, which is why there are so very few of them.

The living conditions of most early humans would not have been so different from those of apes today. Imagine apes inhabiting a jungle; it is a very acidic environment and there is not much water. When the apes die they usually end up on the ground. There, they are scavenged and what is left of them rots relatively quickly because of the acidity of the soil.

The bodies of early humans might either have ended up in a cave or have fallen into a river and been swept downstream into a delta where they would have been covered very quickly and so remained for archaeologists to discover thousands or millions of years later. Those remains were not eaten by predators and were preserved in stable conditions. The human bodies which were exposed to predators would have been consumed from their extremities, as scavengers were keen on the nutrient-rich red marrow which is more easily accessible in the feet and hands. For all these reasons, there is barely any data on the feet of early hominins.

Our earliest ancestors who perhaps walked on two feet are unknown to us. DNA analysis suggests that they existed up to eight million years ago, but there are no fossil remains from which we can piece together a coherent picture of exactly how and when we split from chimpanzees. We know almost nothing about the evolution of chimpanzees. There are only three ancient teeth in the fossil record, and the oldest of those is only half a million years old. But there are older fossil remains that provide a little more background to the human story.

The word "hominin" refers to the lineage of apes that emerged after our species split about 7–8 million years ago, of which *Sahelanthropus* is, to date, the earliest specimen.

Sahelanthropus tchadensis is a 6.7-million-year-old fossil skull discovered in central Africa. Considering it appeared so soon after the species split, it looks remarkably human. It possesses several features of the skull, such as the flatness of the face, and the spinal attachment is underneath rather than at the rear of the skull. Both these features suggest a centre of gravity in the body (for which there are no remains) that is seen in other bipedal hominins.

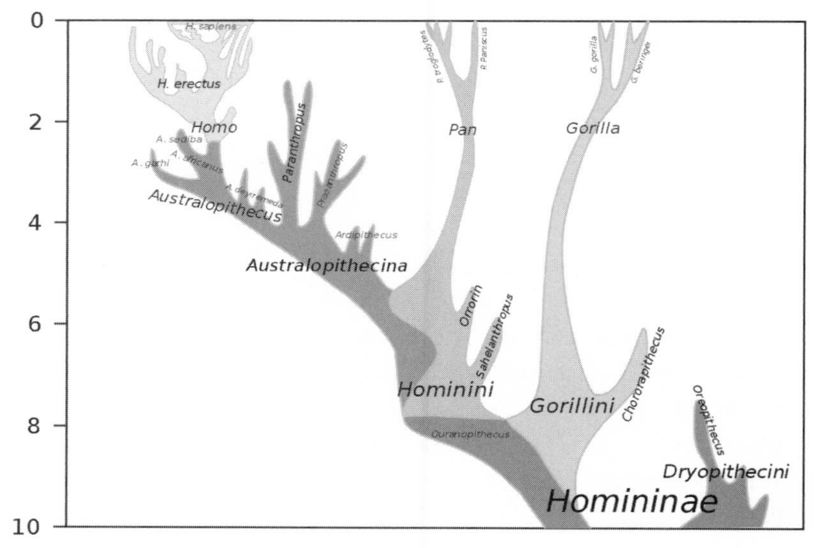

Model of the evolution of hominins over the past 10 million years.

It has other similarities to modern humans, too – its teeth, for example, are quite small and compact.

Orrorin tugenensis (meaning "original man from Tugen") was discovered in western Africa in 2001. The remains are from the same period as *Sahelanthropus,* for which there are 20 fossil fragments. Among them is a jaw, again with teeth similar to those of modern humans. The femur fragment also suggests bipedalism, but the evidence is inconclusive.

Skip forward nearly two million years and there is *Ardipithecus ramidus,* also from western Africa (5.8–4.4 million years ago). One specimen, a 50kg (110lb) female named Ardi, was found with many of her bones intact (and amazingly, even some of her feet). She has very long arms, and a divergent toe, suggesting that she was not an obligate biped (adapted to walk only on two legs), but perhaps a more opportunistic one. *Ardipithecus ramidus* existed for very long time, the second longest of any of the hominin family.

In *Extinct Humans*^ paleoanthropologists Ian Tattersall and Jeffrey H Schwartz recount the discovery of a series of trackways that were discovered at Laetoli, Tanzania, which contain 70 footprints from 3.6 million years ago. These are from a species called *Australopithecus afarensis,* the same species to which the famous "Lucy" belongs. Lucy's fossil remains are staggeringly complete in comparison with others from the fossil record. Forty percent of her remains were dug up, and because most of the human body is built in symmetrical pairs, it has been possible to reconstruct a nearly complete skeleton.

As with other, older remnants, Lucy's feet have not survived, but other *Australopithecus* remains which did have feet were found to have proportions – including the length of tarsals and metatarsals – that tell us they were more similar to chimpanzees than to modern humans (we have shorter toes, which are much more efficient for movement). There are many similarities with modern humans: a more flexible and open lower back, a short and broad pelvis with a large sacrum attached to the spine. Though these early hominins were bipedal, throughout their bodies they retained features that were still suited to the arboreal life of apes. Lucy could walk on two feet, but many aspects of her frame would have made it difficult to do so for sustained periods of time. She had curved bones in her hands and feet; while standing up her fingertips probably reached her knees. The skeletal traits that make up what has been called the full "endurance suite" (an arched foot, enlarged limb joints for better impact resistance, powerful big toe) are only apparent in more recent hominins. At the core of these traits is the technology of the foot.

In most other primates, the foot has a different set of functions and consequently a different design. Apes have prehensile, or grasping, feet. Their big toes are opposable and look a lot like thumbs jutting outward from the central part of the feet. The deep cleft between the big toe and its neighbours makes the foot ideal for gripping while climbing, as well as providing divergent space for substantial muscle development. The phalanges (toe bones) are also curved, enhancing grasping strength. Despite these curved bones, other apes possess flat feet with no arch (plantigrade). Lucy's were something of a bridge between the two.

Homo habilis (handy man[2]) from about 2.3 million years ago, had a smaller jaw and teeth, so could not eat the same diet as its more ape-like cousins (for example, *Paranthropus boisei*), but neither was it an adept hunter with long arms and short legs, and it was most likely an opportunistic scavenger.

Being a land-based bipedal scavenger is about as risky as being a clog-wearing, blindfold tightrope walker with a pride of hungry lions below.

---

2 *Homo habilis* was also an early tool user. Manufacturing tools meant breaking one stone upon another to create a sharp edge that could cut through an animal hide.

On the savannah, *habilis* had a distinct advantage over its feline and canine competitors: its brain. As the guts in these hominins shrank (suggesting an improvement in the quality of their food), it meant that other organs could receive more evolutionary attention in future generations. With greater calorific intake, a bio-structure such as the brain could be more easily sustained. The brain is a hungry organ, accounting for about 20 percent of calorific expenditure, but meat is good at providing that additional nutrition and energy.

The next hominins to emerge from the shade of the trees are thought to have evolved from *Homo habilis* during the period of cooling and drying in which the rainforest retreated, leaving behind savannah or desert. They could also have evolved via a transitional species and are commonly called *Homo erectus*. There is debate about whether the African *Homo ergaster,* meaning "working man", is a different species from the South Asian *Homo erectus*. *Erectus/ergaster* is the first of the human species in the fossil record to show signs of long-distance bipedalism. Bones reflect locomotor patterns, so the length of limb and the sturdiness of the femoral heads are strong indicators of this. *Erectus/ergaster* also

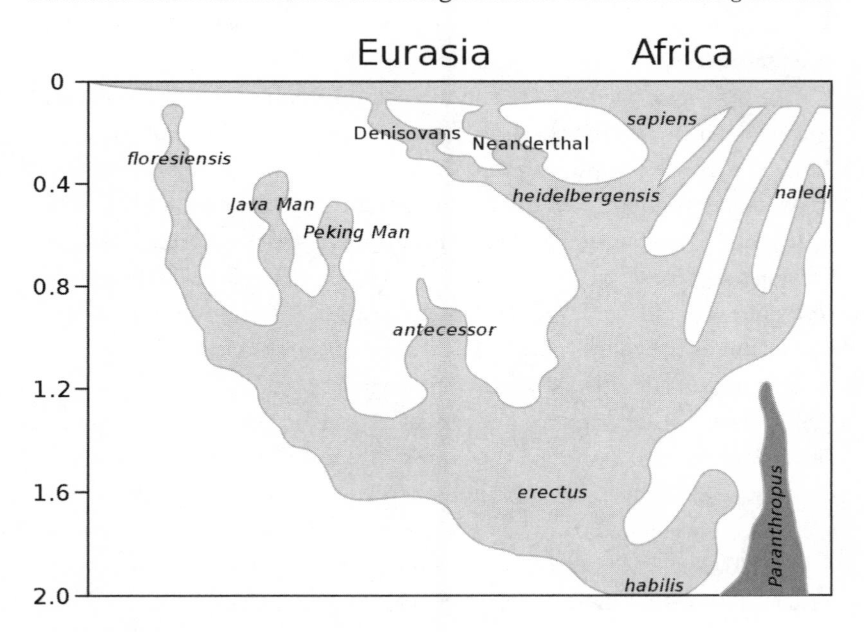

Model of the evolution and migration of hominins over the past 2 million years.

lack quite a few of the features of tree-dwelling species.

A species discovered only in the last few years is *Homo naledi*. These hominins were short in stature. Morphologically, their hands look millions of years old (because they are so curved), but their feet are anatomically similar to those of modern humans. Recent dating, though, has shockingly revealed that they are probably as little as 200,000 years old, with skeletons that are a mosaic of old and new. They had a brain about the size of a 250g (8oz) pack of butter (*Australopithecus* from millions of years before was similar), just over a quarter of the size of Neanderthals (who have the biggest brains of all the *Homo* genus). But *naledi* were able to persist among other hominins in South Africa, and at the same time as *Homo heidelbergensis*. With curved hands like an ape's and human feet, they probably moved between arboreal life and bipedalism.

Homo heidelbergensis is the name for the transitional species between *erectus* and many more modern versions of humans. They emerged about 600,000 years ago, and in their 400,000-year lifespan (longer than ours) they migrated at least as far as northern Europe (the Heidelberg of their name being the spot of their first discovery; *see* map opposite). Their fossil remains have been found throughout Africa.

The best evidence for the locomotive patterns of ancient humans comes not from bones, but from footprints discovered in Kenya in 2009 which were dated to about 1.5 million years ago. This is well over a million years before modern humans walked the planet. More recently, in 2016 a team at the Max Planck Institute for Evolutionary Anthropology and George Washington University concluded that the tracks belonged to *Homo erectus*.^ The footprints revealed substantial similarities to human feet, telling us more about the way they moved than any bones probably could. The gait and biomechanics of *Homo erectus* were much like that of modern humans, with the tracks revealing that the morphology of their foot included a round heel, an arch and a non-divergent longitudinal big toe, like ours. This was a foot that could really move, possessing technology for long-distance travel not seen in other primates.

Being human, it seems, starts in our toes and in the space created where the arch of our foot vaults away from the Earth. More than anything, it is these parts of our anatomy that make us unique. So, make yourself comfortable as we head off on foot to find out more about where our Anthropocene bodies came from, and of the places that they evolved to excel in.

CHAPTER 1

GETTING UP & RUNNING

> Twenty-six bones, thirty-three articulations each with six degrees of freedom of motion, twenty muscles, four layers of arch muscles, a beautiful structure that is meant to be a mobile adaptor, a rigid lever for push off, a base of support, spring-like: all those characteristics and we contain it in a shoe!
>
> Irene Davis

Long after the ink had dried on Charles Darwin's *On the Origin of Species*, the sage of Victorian zoology set to work on understanding aspects of his theory as they related more directly to human evolution: subjects such as sexual selection, evolutionary psychology and race. In *The Descent of Man^*, the result of ten years' head scratching, Darwin explained that he believed we had our feet to thank for our supremacy as a species. The foot freed the hand of the work of locomotion (its most common function in other primates) allowing the hand the opportunity to develop nuanced and highly technical abilities.

Darwin was right to focus on our hands. Their capability and their mechanics are admirable. They have tens of thousands of subcutaneous receptors, and these come in quite a variety, equipping them with the technology to complete highly complex tasks.

Receptor	Function
Merkel cells	pressure, deep touch, detecting shape and edge
Ruffini corpuscles	proprioception, detecting movement on skin (as when dropping something)
Pacinian corpuscles	quick changes in pressure and vibration
Meissner corpuscles	touch sensitive (especially at fingertips and pads), stretching, also vibration, touch and texture; particularly well-developed in Braille readers
Free nerve endings	multiple functions: temperature, pressure, stretching, itching and pain
Muscle spindle fibres	inside hand muscles, detecting length and stretch
Golgi tendon organs	inside tendons of hand, detecting length and stretch

These receptors are all multi-core processors that can process and sort several different kinds of signal simultaneously. For example, if you press a button to call a lift without looking, you immediately feel the pressure of the pushed button and can tell the texture of the button at the same time.

Every second, the thousands of touch receptors in our hands channel information along the peripheral nervous system to the spine, where those signals are zipped up to the brain to be sorted and processed. Digital acuity is hard won because it takes months or years for the brain to strengthen or focus on these signals. But even after a relatively short time, a musician may no longer need visual confirmation of their movements (they don't need to see where their hands are on frets or keys) because there has been sufficient neural remapping for them to know instinctively, proprioceptively, where their hands and fingers are in space.

The evolutionary depth of this system is not known. We might think that the cognitive ability linking creativity and imagination with fine motor control of the hand and fingers and a talent to talk about both of these things all arrived at the same time. Physiologically, those receptors have always been there, just as they have always been there in the foot. But there is a strong likelihood that language first emerged because of the abilities and dexterity of the human hand.

Anyone who has travelled in a country where the language spoken is completely alien to their own will know that there are universal aspects to communication – limited, but universal. I'm not aware of any culture that does not point to indicate, or of one that cannot link the number of fingers in a gesture with an actual number.

Pointing begins in our infancy (usually at about 14 months), but even something so simple as pointing implies a specific world view, as Andrew Whiten argued in his essay, "The evolution of deep social mind in humans".^ To possess a deep social mind is to have a shared idea of the world and an ability in the individual to infer other minds, intentions and beliefs. It is a kind of mind that does not end at the peripheries of the skull but penetrates others. The difference between the way this works in humans as opposed to other primates is that for us it is a fully reflexive system. Apes' impulse to read others' intentions is motivated by scheming and self-interest, so the loop closes at the level of the individual. Humans keep the loop open to allow reciprocation: "I read your intentions and also signal my own back to you." This is the essence of human communication and community.

This experience of a shared world view is linked with a theory about the emergence of language itself: that words began as sounds linked with gestures. Hand gestures persist as an integral part of everyday speech. They are so closely linked that the regions in the brain that process hand and mouth movements are next to one another in the cortex. In infancy, gestures are the principal mode of communication and they later become supplementary.

Gene expression is important, too. One of the most important of these is the FOXP2 gene, which among other things seems essential for language and makes it easier for us to change new experiences into routine procedures (such as speech, or playing a musical instrument). What the FOXP2 gene does is encode proteins whose function is to activate certain genes in the brain. In humans it carries two mutations. Those mutations have been reproduced in mice and the effect was a marked improvement in certain motor abilities (solving a maze, for example) and vocalization (the mice were more inclined to use vocal signals). Although the FOXP2 gene is shared with Neanderthals, and even though they had the anatomy and morphology to enable speech, whether they actually spoke is unknown. The evidence suggests that it is possible, even likely.

There are several theories as to why language emerged. One of the least persuasive is that hand and body gestures predominated as modes of communication until tool use developed to the point when the hands were rarely free to signal. Or that speech emerged as a technology that could penetrate darkness or travel longer distances. Others have linked it to our eyes. Because our sclera (the whites of our eyes) are clearly visible, our direction of gaze can be shared in ways that are not possible for gorillas (whose whites, like those of most mammals, are masked). It is only easy to see which way a gorilla is facing, not what they are looking at. Being able to see the eyes of another human, to read their thoughts and emotions in them, became so key during our early evolution that those who did not have this physiology or ability either died or were killed before they could reproduce. The ability to read minds was essential for early human survival.

Further links between language and gesture emerged with the discovery of Broca's area, a part of the brain in the frontal lobe of the dominant hemisphere which is activated both when we gesture and when we speak.

After several centuries of debate, we are not that much closer to discovering when humans first spoke. With no empirical evidence, we are left wandering amid theories which attest to the fact that the origin of language is one of the toughest problems in science because there is no discoverable data.

Archaeology tells us that Neanderthals had the right kind of morphology to produce verbal language and that *Homo heidelbergensis* were likely to have used forms of vocalization (like other primates); we can infer that as their cultural lives developed, their language did, too – but there's no proof.

We know from studies and stories of feral children (from Kaspar Hauser at the beginning of the 19th century^ to Genie the socially-isolated and abused child who was rescued by authorities in California in 1970) that modern humans acquire language and complex grammar acquisition at a particular time (or "critical period") in their lives, in the same way as height or jaw development. Once the opportunity passes, it is gone forever. In the case of Genie, who was rescued at the age of 13, she had been so neglected and deliberately isolated that she never learned to develop language. She acquired a wide vocabulary of words, but never mastered grammar or syntax. If language acquisition doesn't happen during childhood, it looks as though it cannot fully develop in adulthood.

Language acquisition seems to be an unbelievably sensitive Jenga tower that relies on all the blocks being in exactly the right place to keep the structure upright. Language in deep history had to develop over numerous generations and the morphology had to remain solidly in place from generation to generation or the link would have been permanently broken. Language, even in its very simplest forms, must have been evolutionarily viable and advantageous from the start for it to have been selected for so strongly. The history of our hands and our speech is buried here in a past dominated by a network of lost languages and probably many, many thousands of false starts by minds that ventured outward through speech into the world but found no other mind to recognize and no one to hear their thoughts.

While our hands are praised richly for their abilities of touch, manipulation and creation, our feet are also highly technical, with some 200,000 subcutaneous receptors. Although modern humans lace them up in 25cm (10in) coffins, changing their shape and function, making them weaker and less effective. To provide some context, think of the amazing things your brain is able to do with the information it receives from the central trichromatic photoreceptors in your eyes.[3] There, light passes through the cornea, lens and other parts of the eye to land on one of three (or four in some women) different kinds of receptor whose molecules jiggle when they recognize certain wavelengths of light. That jiggle is converted to an electrical impulse that emerges from the other end of the receptor into a mesh of nerve cells which eventually bundle into the optic nerve, where it is sent to the visual cortex of the brain to be processed. The visual acuity we share is founded upon a similar number of just 200,000 receptors, suggesting that evolution had more ambitious plans for our feet than ensconcing them in shoes.

About a third of our brain's operative power is given over to processing visual stimuli from those receptors; but our feet, coddled as they have been for so long, have lost their sensory language because they receive so little stimulation from the world. In terms of being a barrier to our senses, you could say that shoes are to the feet what a blindfold would be to the eyes.

If Darwin is correct, and he frequently is, bipedalism also made it easier for us to carry items such as tools, food or our young. Being on two legs rather than four might have made evading predators easier, too, and it certainly helped us travel to and from important resources such as food, water and shelter more easily. The earliest species that ventured out of the pond had to stay within a few feet of water; but when bipedalism became obligatory, humans were free to wander many miles from it, and eventually to walk across and between continents with free hands and strong feet.

The story of our feet and of the muscles in our backsides that allow us to stand upright on them, are the focal points of the beginning of our journey

3 These are the colour receptors in our eyes usually called cone cells. Most humans have three different kinds (bees have three, too). Fish and birds have four different kinds, but many other mammals, such as dogs, ferrets and hyenas, have only two.

toward the Anthropocene body; and how the changes we made in the world ended up changing us.

FREE HANDS & BIPEDALISM

The fact that the feet of early hominins freed their hands would have been of little account unless the ability was paired with the talent to do something with them. It is not easy to say at exactly what point in evolutionary history manual dexterity emerged. We have been on our feet for well over two million years, but that does not mean that our hands were of great use to us.

Research published in the *Philosophical Transactions of the Royal Society* in 2013^ suggests that bipedalism and manual dexterity were formed independently of one another. There is a word for the point to point mapping between parts of our body and parts of our brain and nervous system: somatotopy.

The somatotopic representation of our fingers in our nervous system is what allows us to play a Schubert piano sonata; to do so, the digits must be individually mapped so that they can move independently of one another (they also allow us to type on a computer keyboard, for example). While monkeys also share some of this potential dexterity, that does not mean that they are able to deploy it in the same way: with different bodies, different environments and different priorities their dexterity did not lead them down the same evolutionary path.

There is no evidence to show that a part of the genus was using a sharpened spear like a javelin until *Homo heidelbergensis*. More than another hundred thousand years went by before Neanderthals used spears with sharpened stones (in the shape of leaves) at their points – and it was another quarter of a million years before those sharpened stones were made into arrow tips.

For the majority of our past a spear and a bow and arrow did the work for us, but what use were they if you couldn't get within throwing distance of a quarry?

The feet of a monkey might resemble their hands, but do not possess much ability to manipulate. The foot is good for grasping and brute strength, but not a lot else. Anyone who has tried to write or paint with their foot might believe that humans are the same, but we are not. While we cannot play the piano with our toes,[4] we do have some independence of movement in our extremities.

In humans, the central triad of toes are mapped together, with the little toe at the end retaining only the most modest kind of self-government (a bit like Gibraltar). A sizeable difference lies in the big toe (the hallux, from the Greek for "I leap") which has a separate function from the other toes. It is strong, can flex fully independently of the other toes, and has muscle attachments that connect the foot to the lower leg, suggesting a strong link with locomotive bipedalism.

FOOT TECHNOLOGY & HUNTING

The human foot has four layers of arch muscles supported by strong ligaments running along the underside of the foot. Not only do these muscles prevent the arch collapsing, they also store propulsive energy which can be loaded with body weight and return that energy on toe-off when walking and running. An opposable hallux would squander this stored energy, whereas the human foot can capitalize on it.

The precise shape and function of the human foot maximises in every way its potential for movement. The fact that all the toes are longitudinal means that the energy stored in the arch during the first half of the gait cycle can be returned, aiding propulsion in the second. The longitudinal hallux returns that final bit of extra energy in human walking or running, providing a little spring to every step. It is a truly ingenious mechanism that uses the body's momentum and weight to create locomotive energy.

Homo erectus seems to have had a modern foot for at least 1.6 million years. One set of remains, called Turkana Boy, are remarkably complete. His body implies countless adaptations for endurance running: long and strong legs, large muscle attachments suggesting sizeable glutes (gluteus maximus, the main muscle of the hip), a nuchal ligament to stop the head pitching forward (as it does in modern humans), smaller shoulders and a spine that is able to counter-rotate (from that exposed lumbar region). The connection between the pelvis and the spine is more robust in *Homo erectus* than in other ancestors, meaning better shock absorption and stability (this also goes for hips, knees and ankle joints). Even our efficient and enlarged

4 As in the episode of *Tom and Jerry* in which the cat bangs out a piano concerto with his claws.

stability (vestibular) system, that allows us to focus on something while in motion (as well as process motion, equilibrium and spatial orientation data) is also found in *Homo erectus.* As in other humans, this sophisticated balance-and-focus system is larger than in other primates.

The early human body is fractal in its brilliance; the closer one looks at it the more its ingenuity becomes apparent.

These adaptations met an environment that was changing, in which the body became almost perfectly matched to its place. About 1.8 million years ago there was an environmental shift in east Africa from more closed habitats to open vegetation: the savannah that we recognize today. (Even our height was perfect for peering over the tall grass.)

Homo erectus possessed a strong talus (ankle bone), stronger than their ancestors', a longitudinal arch and – despite the high number of moving parts – an ability to convert them into a stiff lever that could transfer power from the calf muscles near the knee all the way to the tips of the toes. Other research by paleoanthropologists (particularly Daniel Lieberman and Dennis Bramble) has also highlighted the importance of, not just the morphology of the foot, but of the entire body, to the ability to run.^

Persistence hunting, a mode of stalking and killing in which the participant(s) chases the quarry to the point at which it collapses, made early humans the most fearsome hunters on the planet.

Running speed was less important than one might imagine. The longer leg and shorter arm length of *Homo erectus* meant that they could probably run faster than later *sapiens* (namely us), their long shanks giving them a particular advantage (as elite west African runners have today). But an ability to sprint is not that much use to humans, except in emergencies. It's stamina we need. While practically every quadruped on Earth could – and still can – outrun us over a short distance (including rabbits and squirrels), our cooling technology (perspiration) was incredibly effective, so after several rounds and recoveries, we could catch up with practically any animal, as long as it could not clamber to places inaccessible to us.

Being on two feet also meant that a lot less of the day's hottest sun hit our bodies (about 30 percent less than a quadruped), so not only were we better at shedding heat, we also absorbed quite a bit less of it in the first place. We had the hypothermic advantage: in a chase, we were less likely to collapse from heat exhaustion than our prey. Our ability to run was important, but it would have been useless without our cooling system.

While previous versions of our feet (in preceding species such as Lucy's *Australopithecus afarensis*) tell us about our need to negotiate branches and trees, or our need both to climb and walk, later hominins became increasingly adapted for living on the plain.

Genetic research is beginning to show that there are implied species families that there is, as yet, no evidence for in the fossil record. One version of this species is called *Homo antecessor*, one speculated to be between *erectus* and *heidelbergensis*, or the last common ancestor of Neanderthals and modern humans. But with only a few fragments of bones, and those probably belonging to children, there is not much that can be garnered from them. Footprints (the largest a UK size 8) thought to belong to *Homo antecessor* were found on a Norfolk beach in 2013, and these were dated to more than 800,000 years ago – a time when the climate would have resembled that of Sweden or Norway today. The footprints (where the arch is visible and some of the toes) suggest that the group were either barefoot or had very meagre foot coverings that allowed a recognizable impression to be left in the sediment.

These were feet that had migrated all the way from Africa and just as it is wrong to assume that there is a nice linear relationship between the species, or in the evolution of parts of the body, it is equally wrong to assume that migration worked in a linear way too. The genetic research shows that we diverged from Neanderthals and Denisovans (half a million and a million years ago). Tests analyzing the genetic structure of Denisovans reveal that it differed from that of modern humans in 385 nucleotides (the basic structural unit of DNA). To make sense of that difference, there are only 202 differences between us and Neanderthals, and more than 1,400 between us and chimpanzees. The DNA sequencing also suggests that modern humans mated with Neanderthals and Denisovans on numerous occasions.

We only have a tiny part of the story, but the science seems to be gathering pace with every week that goes by. Galloping at the speed we once did about the plains of history, chasing down prey with our sticks and stones, then feasting for days on what we caught, cooking over our recently discovered fire – happy as the proverbial Larry.

See the graceful ease with which nostalgia creeps in?

OUT OF & INTO AFRICA – WALKING & THE EARLIEST FOOTWEAR

The discovery of fire is a hot topic among paleoanthropologists. Experts cannot agree who started it, how it spread or what it was used for. But we know that the kinds of migration that happened in the Pleistocene would not have been possible without the use and control of fire. The earliest evidence for it is sketchy, but it looks as though *Homo erectus* may have been able to make fires up to 1.5 million years ago. Evidence for cooking with fire dates from much later. A cave in South Africa has revealed evidence of carbonized leaf and twig fragments which date from about a million years ago.

Fire allowed our days to lengthen deep into the darkness, provided us with greater protection from predators, deterred many insects and pests – and by allowing humans access to better nutrition it probably enabled them to grow a little taller (cooking, as we Anthropocene humans know only too well, makes it easier to consume both macronutrients and micronutrients). Fire also made it easier for species to move and migrate more freely into colder climes, because cooking and drying foods enhanced both their preservation and their portability. Instead of setting up home and hearth, fire probably made early humans more, not less, mobile – able to travel long distances over long periods: to migrate.

The ways in which fire can be said to have changed our bodies will be the meat of other chapters. Exactly how it might have changed our feet is a little more difficult to say. (The precise foot sizes of *Homo erectus* are unknown, but if their height is anything to go by – rarely cresting 1.5m (5ft) – they were unlikely to have challenged Sasquatch.)

How did they hunt and migrate barefoot? Well, the pads on the sole of the foot are designed specifically to adjust to tougher surfaces. The hard skin that develops there is the foot's adaptive response to increased friction and pressure. It creates a much tougher layer of keratin to protect the dermal layers beneath. The skin layer (as those of you who have taken a blade to it will know) is completely dead, made up as it is of undifferentiated cells in a strong adhesive network. Yet another link with the limbs of a more primitive common ancestor is that keratin is the same stuff that hair and horns are made from. Walk over a rocky and jagged pass and you are basically developing the hooves of a mountain goat.

While our feet might share some of the qualities of hooves, anyone who has ever stepped on an upturned plug will know that they can hurt like hell. And if we are reading the fossil record correctly, there must have been a glut of upturned plugs scattered about the plains some 40,000 years ago.

Early human migration is not just about being able to move over terrain; it is also about climate. Evidence for the migration of modern hominins out of Africa currently dates back about 200,000 years, but at the rate these discoveries are currently being challenged, it will probably soon be found to have been earlier than this.

This was the period in which the climate was still volatile. The last Ice Age had been so severe that it was thought to have threatened an extinction event, which for a number of species, it did.

Homo heidelbergensis, the descendant of *Homo erectus*, flourished in Africa and Europe for hundreds of thousands of years. They outperformed *Homo erectus*, perhaps with enhanced intelligence which gave them control of fire, more sophisticated tools and with evidence of complex behaviours such as collaborative hunting it has been argued that they may have possessed some language. About half a million years ago the species spread throughout Africa, the Middle East and Europe (remains have even been found in West Sussex on the south coast of England). They were the first in their genus to build shelters and cook on a hearth.

Throughout this period the climate was changing rapidly, with the ice caps advancing and retreating in a seemingly unending cycle. During cold periods northern Europe became a sugar-white tundra, vast and barren. During the thaw, forests of beech and birch proliferated. *Heidelbergensis* adapted to this changing climate as best it could – compared with its African cousins, it became shorter, stockier and more robust; this was a body better adapted to retaining heat. Eventually, it became *Homo neanderthalensis*.

The Neanderthals were better built to survive in the changing climate. The harshness of it can only be imagined, with blankets of ice stretching for hundreds of kilometres. The ice sheets over what is now Britain would have been hundreds of metres thick, taller than any building in the UK. The Neanderthals survived on the fringes, migrating with the ice as it retreated or advanced.

The other family of humans that emerged from *heidelbergensis* are the Denisovans. While the Neanderthals were found in western Eurasia, the scant evidence we have for the Denisovans has been found to the east – the only fossils are a couple of teeth and a finger bone. Denisovan DNA, sourced from that bone found in the Denisova Cave in Siberia, is distinct from both Neanderthals and modern humans. The size and shape of their bodies can only be guessed at.

There are, though, numerous Neanderthal remains. DNA analysis indicates that some had pale skin (better adapted to convert limited sunlight into vitamin D), and also red and blond hair. On their meat-rich diet, their brains became the largest of any humans (about 10 percent bigger than ours). And apart from the fact that no one has been able to overhear a Neanderthal conversation, practically all other evidence suggests that they were able to speak. They hunted at the peripheries of forests, preying upon large mammals such as deer. They were good stalkers with an array of weapons, but when the big freeze came in waves between 50,000 and 40,000 years ago and a colossal ice sheet covered northern Europe, freezing much of the north Atlantic, the Neanderthals did not survive. Evidence of cannibalism in some fossil remains imply practices or customs unknown to us or extreme desperation as everything around them turned to ice.

During this period it's estimated that the human world population dropped as low as 10,000. If one of the cold snaps had been a little more intense, you would not be reading this (and equally there would be no such thing as reading).

While there was a good deal of interbreeding between modern humans and Neanderthals, their different bodies meant that their lived experiences were equally divergent. Their sturdiness would have made them less well adapted for long-distance walking and running. Though they would have had "modern" feet they would not have been light on them. It is estimated that the men burned more than 4,000 calories a day in their constant hunt for food (a heavier body requires more sustenance) and their efforts to keep warm.

To claim that the story of our species' migration is slightly complicated would elicit a belly laugh from anyone working in the field of genetic paleoanthropology. The map of human migration used to look a little like the opening credits to the sitcom *Dad's Army*, showing confidently thick arrows and clean lines of linear movement. As more is learned about how DNA was shared among different hominins, the picture is becoming more complicated, with histories of interbreeding and implied species stretching back hundreds of thousands of years. If the very first migration events (such as that of *Homo erectus* into southern Eurasia) might be seen as a ripple from a single pebble thrown into a pond, then migrations since then are more like the ripples created by flinging a huge handful of gravel into the pond. Genetics and history both tell us that once we got up on our feet we moved and migrated in all directions practically all the time, having quite a lot of sex either with one another, or indeed with any species of *Homo* that might accommodate us along the way.

En route, tool use developed. Sharp projectiles appeared (probably an invention of *Homo heidelbergensis* about 400,000 years ago). The Neanderthals struck on a new method for knapping flint 150,000 years later. The Levallois technique for shaping flints meant that they could make lighter and sharper tools and hunt larger animals more easily.[5] The abundance of supply contributed to population growth, as did the growing sophistication of technology and trade (the earliest evidence for which appears about 150,000 years ago) – long, long before documented history. The earliest known evidence of personal adornment appears around the same time, with the shells of sea snails being used as beads.

A little later again came the technology of shoes. The kind of footwear invented for warmth – a furry wrap made from animal skin – is probably in the region of half a million years old. Such a soft foot covering would have had no impact on foot strength, bone density or development. (Soft and unsupportive shoes still make the foot do all the work and therefore stimulate bone density, muscle growth and strength.) These sorts of soft wraps would also have left practically no trace in the fossil record.

Instead we can see hard and protective footwear emerging in the later Upper Palaeolithic period. In *neanderthalensis* and *sapiens* fossil remains, the four smaller toes are less well developed than in earlier records.

Shoes changed the way we moved, but they also allowed us to travel farther on foot in harsh climates. Over a number of years, these changes to our gait became mapped onto our bones. Footwear that pushed against our feet caused changes in gait, which altered the connection between many of our moving parts. The difference is easy to imagine when we think about the way we move in a pair of slippers compared to some shoes worn to work, or 15cm (6in) high heels. The shoes flattened our arches and weakened intrinsic muscles as a result of disuse or misuse, as well as causing more serious problems such as bunions, in which the bone itself becomes inflamed.

---

5 The technique (named after the French suburb in which a stash of such tools was discovered in the 19th century) involved the creation of tools by cracking long and fine chunks of flint from a base stone. Because the stone breaks in large flakes, the tools are comparably light and sharp.

Physical anthropologist Erik Trinkaus reported his study of these fossil remains in the *Journal of Archaeological Science* in 2005^, comparing the formation and development of their phalanges and discovering a relatively sudden change between the remains from the Middle and Upper Palaeolithic (between 26,000 and 40,000 years ago), when toe bones became less developed. After studying the relative chunkiness of early Native American toes (those of people known to have walked barefoot) compared with those of some Inuit (wearers of heavy sealskin boots), he concluded that the shod foot is not exposed to the same levels of stress as bare feet, and consequently thins out.[6]

Shoes, it seems, were part of the creative explosion that emerged about 40,000 years ago when Shamanism, ritual and religion, and cave art began to appear. It seems inevitable that this period should also have seen the refinement of footwear.

If the daily estimate for early human mileage is correct, at 8–14km (5–9 miles) a day, then it did not take many generations for a species to travel to every part of the habitable and accessible globe. The recent discovery of an upper jaw bone belonging to a *Homo sapiens* in Misliya, Israel, has shown that modern humans were on the move much earlier than previously believed (by about 140,000 years). A trove of *sapiens* teeth has also been discovered in China and they have been dated at about 100,000 years old.

From the very beginnings of our species, migration and movement seem to have been as natural to us as enjoying a really good sunset or some mammoth steak.

6 Our bones are always in the process of making and remaking themselves to cope with the stresses we place on them. This is why the bones of runners – who repeatedly pound their body weight through their skeleton – have greater density than, say, a swimmer's or cyclist's, whose body weight is supported during exercise.

CHAPTER 2

STAND UP

Everything needs to move.

Gary Ward. *Anatomy in Motion*

Making sense of the length of time *Homo erectus* spent running around the African savannah is beyond the capacity of the human brain. In the context of our own lifetime, 1.9 million years is just too great a multiple of it for the number to make any sense to us. For this unimaginably long period these grasslands were home to the millions of early humans who existed on them. For those changing bodies all that time ago, the savannah was the ideal environment. And unlike the other landscapes and environments we are to visit in this book, it was the one in which humans appeared, but it was not *made* by them.

The early human body and the savannah only met by chance; it was a one-in-a-billion fluke in which the adaptations the hominin body was undergoing suddenly made it ideal for the time and place it found itself in.

But it was not really a fluke at all. Primate history is littered with individuals and species born in places that later changed, rendering calories and nutrients more difficult to come by. *Paranthropus* is one example. There are countless differences between *Paranthropus* and *Homo erectus*, but one of the most noticeable is their comparative size. *Homo erectus* was both taller and thinner, but even so, the relative gut size of the two species show evidence that the more diminutive *Paranthropus* possessed much larger and longer guts than *Homo erectus*.

As the *Homo erectus* body emerged, it showed selections for smaller and shorter guts, suggesting that the quality of its food had improved: either its body was better at processing nutrients from food, or that processing was done outside the mouth (through cooking, for example). The evolutionary selections suggest that as agility was optimized, *Homo erectus* ate better food as a result, leading to further selections in a rising spiral which optimized agility.

At the core of all this agility and power is a muscle group not unique among primates, although the shape, size and functions of these muscles are. The fossil remains of early humans show substantial muscle attachments on the rear of the pelvis, indicating that *Homo erectus* possessed big ... bums. These huge muscles are as important for stalking and hunting as we might expect any weapon to be.

THE ANATOMY OF HUNTING

Homo erectus means "upright man", and without our gluteal muscles we could not have stood on two feet, let alone have covered billions of miles of the savannah – and later, Africa and Southern Eurasia.

To say that our bodies needed particular attributes to be able to hunt is to underestimate the combined role that different abilities and specialities played in stalking prey. Everything from the tips of their propulsive toes to the hair that shaded the tops of their heads from the sun was key to making *Homo erectus* good hunters. Joints needed to be able to absorb impact; their bodies needed to be able to shift enormous amounts of heat from the exertion of the hunt.

Homo erectus developed a particularly good throwing action. But what use was throwing if their eyes couldn't stay focused on a quarry while in motion? There were at least a hundred more facets and features that made them good hunters, and in a chart that ranked their importance, the gluteal muscles would probably be near the top, with their senses and the technology of their feet bunched close by.

The gluteals (that give us our distinctive bum shape) are a sextet of muscles, the larger of which are strong, tough, coarse and almost injury proof. Most can lift perhaps a couple of hundred pounds with them. Although we stand on our feet, it is these huge muscles behind the central hinge of our bodies between our back and legs that really do the work. Without them, we would topple forward onto all fours.

There are three layers of these six muscles in each cheek. They are: the gluteus *minimus* (the smallest and deepest of the triad), the *medius* (the middle one) and the *maximus* (the largest one that does most of the work and gives humans their distinctive appearance). It is this last one that is really unusual. It performs a wide range of functions, including extending the hips, hip abduction (moving the leg out to the side), pelvic tilting, trunk extension and rotation. In this tough muscle the fibres run diagonally from the pelvis to their attachment at the top of the leg.

Apes also have six gluteal muscles (three on each side), but for them, the outer one is not called the *maximus* for the simple reason that it is not the biggest muscle in their body. It is called the *superficialis*, which means that it is the outermost of the three muscles. Its size is insignificant in comparison with its two cousins. Our *maximus* needs to be *maximus* because it performs functions specific to human movement: walking and running.

The *maximus* has two main functions; first is pelvic stabilization, keeping the pelvis centred above a hip joint which has a lot more work to do in a biped than in a quadruped. The second is as a hip extensor; it straightens out the joint. The work of the hip extensor is particularly important in running. It is also key in walking and standing upright. It is the *maximus* that is the most "human" of the three. The other two have slightly different functions.

The *medius* is also a substantial muscle, thick and short, built for power rather than speed. It maintains side-to-side balance, which is particularly

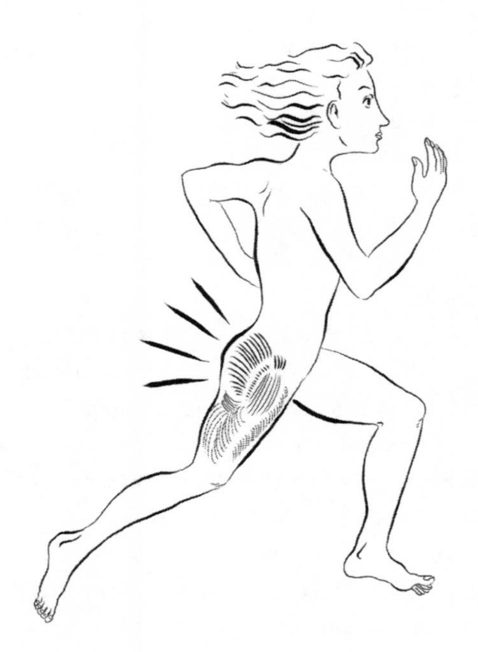

Much of the propulsive power of running comes from the gluteal muscles.

important while walking. The medius is also a powerful abductor.[1] A misfiring *medius* can lead to something called lateral pelvic tilt or a Trendelenburg gait.

When most people walk the pelvis stays almost parallel with the ground when viewed from behind their belt line. But lateral pelvic tilt is recognizable in some people's gait. As you look at them from behind, their belt seesaws from side to side like a ship in a stormy sea – and this can lead to all sorts of problems, especially lower back pain. The causes and associations are many, and most are connected to modern life. Sitting for long periods on a fat wallet (not a habit that troubled early *Homo*), side sleeping, which shortens one side of the body (and is only really comfortable on a mattress), driving for long periods in a fixed pose (one arm out of the window) and carrying a bag on one shoulder all encourage the body to alter its mechanics by allowing it to adjust to its lopsided position.

The *minimus* acts as a back-up for much of the work of the *medius*, assisting with abduction and stabilization when the weight is on one leg.

The glutes are a land-based muscle group. The posterior muscles in the legs of apes tend to be short and very powerful, good for generating lots of propulsion and acceleration over the full range of motion. Basically, they are great for climbing. Human biomechanics trade off this power against a substantial gain in efficiency, specifically over longer distances.

On the savannah, the glutes got a full workout. During the Pleistocene and Holocene, humans spent much of their time walking and running and this meant work for the glutes – and although *Homo erectus* might have lain down to rest, another very common pose was the squat. The deep squat was the ancient equivalent of a bit of a sit down. It was comfortable and the pose could be deployed for relatively long periods.

THE PRICE OF A LITTLE SIT DOWN

Wolff's Law tells us that our bones respond to the loads under which they are placed by remodelling to become either stronger or weaker over a period of time.^ The bones become stronger when they are placed under stress or repeated loading. The reason they become weaker is because the body will not waste calories supporting a dense skeletal system when it

1 Abduction is a movement that is the first part of a sideways karate kick. Adduction – returning the leg to underneath the body – is the second half.

is not necessary.

The equivalent for soft tissues is Davis's Law.^ Here it is, used for the first time in John Nutt's 1913 study *Diseases and Deformities of the Foot*.

> Ligaments, or any soft tissue, when put under even a moderate degree of tension, if that tension is unremitting, will elongate by the addition of new material; on the contrary, when ligaments, or rather soft tissues, remain uninterruptedly in a loose or lax state, they will gradually shorten, as the effete material is removed, until they come to maintain the same relation to the bony structures with which they are united that they did before their shortening. Nature never wastes her time and material in maintaining a muscle or ligament at its original length when the distance between their points of origin and insertion is for any considerable time, without interruption, shortened.

This is an important theory about what happens to our bodies, both within our lifetimes and as the landscapes around us shift and change. Wolff's Law means "use your bone density or lose it" and Davis's Law means the same thing for our soft tissues. But the implications of the law explain more about what happens to us in modern life than a bit of bicep shrinkage. The law also says, use your *range of motion* or lose it; use the capacity to store energy in your ligaments and tendons or lose it.

So much of what happens in the Anthropocene body is about the loss of different ranges of motion, and those losses are compounded by the body's attempt to recruit other joints to compensate, or by acquiring a similar loss.

The most common reclining pose in modern life is to sit in a chair, but our species family didn't sit down in chairs for more than two million years (so chairs will appear later in our story). There were no chairs when hunting, and hunting was what this early body was about – hunting and, of course, eating. The latter was regularly accomplished in a squat. In modern life, the idea that the full squat is a comfortable way to rest or eat seems laughable to the billions who prefer to sit.

In a full squat, our early ancestors would have used their pelvis as a lever. Although it is more complicated than this, they needed to angle their femurs (upper legs) against their bodies at about 10–15 degrees (the angle of a clock hand between 11 and 12 on a clock face) and keep their centre of gravity squarely above their feet so that their weight was evenly spread across the heel and the forefoot.

This resting squat also opens up the lower back (bliss for some back pain sufferers), but to do this the human needs to have long calf muscles. And, because Anthropocene humans have spent much time in shoes, heels or sitting in chairs, their calf muscles have responded by shortening, making it harder to flatten the foot with even weight distribution *and* a gentle forward tilt to allow a resting position with a good centre of gravity.

I would guess that fewer than one percent of adults in the US – or the UK, Europe, Russia and Australia – are able to rest comfortably in a squat for three minutes. In India before colonization, the most common reclining pose was sitting cross-legged or in a squat, but most of the younger generation, having attended schools and universities and worked in sedentary jobs, can no longer do this. Like many of us, they lose their range of motion mid-childhood (they are trained out of it, an idea we will return to later, *see* page 148). But in Indonesia and other parts of southeast Asia locals regularly rest like this, in squares or by the roadside, just for the pleasure of watching the world go by. The Indonesian national pastime is jokingly referred to as *duduk-duduk* (sitting), second only to *rokok rokok* (smoking).

A resting squat, where the centre of gravity is evenly balanced across the fore- and rear-foot, and the weight is planted deep into the hips.

WEIGHT-BEARING FASCIA

We have always been a species that loves comfort. In fact, our problem-solving skills and our capacity to apply technological solutions were put to early use in the most primitive ways to enhance easeful repose. But even a quick survey of the skin types on our bodies will help to provide some indication that our bums are not physiologically adapted to sitting for extended periods.

The skin on our hands and feet is different from anywhere else on our bodies. First, it has an extra layer, the *stratum lucidum,* which helps protect the other layers from the effects of friction. The skin on the ear lobe is soft and velvety by comparison.[2] The skin on our arms is flexible and mobile. The skin on our legs is hairy, but on our extremities it's different. There is no other site on our bodies that possesses the whorls, ridges and loops that our hands and feet have.[3]

People with darker skin have observably lower levels of melanin on their soles and palms. Melanin is not needed to block ultraviolet light on our weight-bearing fascia because these parts of our bodies are expected to be ventral (meaning underneath, the opposite of dorsal). They are meant for base support, especially during locomotion.

Our weight-bearing fascia then, have multiple similarities. They share a slim layer of protective fat, skin that responds quickly to stimulation (hardening or softening as necessary) and is anchored to the bone beneath. Subcutaneous receptors are densely packed in both hands and feet, and they feed us all kinds of data for effective movement and spatial perception. These weight-bearing fascia are sensitive as a rom-com, tough as an action movie and as hardy as a horror movie villain who just keeps getting back up.

But not the glutes, their time should be spent predominantly out of contact with any surface – that's not what they're for. They're meant for work. But how much?

---

2 Quite why we have ear lobes at all is a mystery – it is probably a secondary effect of a gene that is also active elsewhere in the body.

3 Here again, no one really knows why we have whorls, ridges and loops on our hands and feet. Possible reasons might be that they enhance grip at higher intensities, or increase digital sensitivity, allowing for greater nuance in finer tasks.

A "NORMAL" PREHISTORIC WORKING DAY

For decades paleoanthropologists have been trying to calculate how many hours a day the hunter-gatherers of prehistory worked. Any calculation can only be approximate at best, but the numbers are still surprising.

Hunter-gathering requires the refinement of a range of skills, from understanding which plants and berries are edible and which poisonous; which nuts are easiest to crack; how to track and pursue a range of wild animals; how to endurance hunt, how to make and use tools; how to prepare meat once caught; how to build shelters, make and manage fire and clothes; how to protect offspring; how to cook – the list goes on. In turn, the body that evolved during the Pleistocene epoch adapted to meet those needs and that environment. But did *Homo habilis*, *erectus* and *heidelbergensis* slam down a solid 60–70-hour week because they had so little technology to help them achieve their goals?

In the 1960s anthropologist and ethnographer Marshall Sahlins suggested that these were the first affluent societies, in which people were able to enjoy the luxury of all their needs being regularly met.^ Sahlins believed that for these early societies to fulfil their needs they would have had to work a 3–5-hour day.

Since the sixties, Sahlins' ideas have come under much scrutiny by others, particularly the idea of "affluence". The length of the working day has not changed all that much, only rising in more recent estimates to about six hours. That leaves a great deal of time to be accounted for. Hanging out, gossiping, watching the world go by are all activities that early humans greedily indulged in, but they all sound rather boring. There has even been the suggestion that what drove human innovation was exactly this: boredom.^

The bodies of those later humans who had migrated from Africa and deep into Eurasia and colder northerly climes would have had to work hard to subsist. They would have had to travel farther to find the sparser food that the environment offered.

Fossil remains show that even with these seemingly diminished hours of work, their bodies worked hard. Comparative bone density studies of hunter-gatherer and agriculturalist specimens demonstrate that once humans started farming their bone density dropped dramatically. The research was the first that scanned bone density from specimens spanning several million years and it concluded, "density remained high throughout human evolution until it decreased significantly in recent modern humans, suggesting a possible link between changes in our skeleton and increased sedentism."^When we changed from being nomadic to settling and farming, our bone density dropped.

What is very easy to miss here is that the bone density of those ancient farmers was still *so much* more developed than ours is today. Research published in November 2017 showed that while the bones of the 94 women scanned showed variations in strength, their specimens from between 5300BCE and 100CE had on average bones that were 30 percent stronger than those of non-athletic modern women.^ These early farming women worked hard, and it showed in their bones, putting paid to those locker-room assumptions that women tended the home while men did the hard work. The results of this trial also suggest that these specimens had worked harder than today's Olympian rowing teams and their "humeral rigidity exceeded that of living athletes for the first ~5500 years of farming."

Even though bone density dropped when nomadic hunters became settled farmers, the bones of these sedentary dwellers were still significantly stronger than those of modern-day Olympians. Hunter-gatherers would have been stronger and fitter than most of us Anthropocene humans can even imagine.

In deep history, the working day might not have been long, but human bodies show us that it was certainly hard. The bones show that there was a wide range of work and playful activity. Several factors contribute to bone density, some of which we will return to when we crest the Industrial Revolution, but one of the key stimulants is loading. When a bone is put under stress osteoclasts inside it go to work rebuilding and strengthening it. Many activities contribute to bone density, but especially heavy work, such as making flint tools or scraping down a hide so that it could be worn. These were tough jobs, especially in comparison with today's standard work paradigm: depressing buttons on a keyboard to a depth of 4mm (0.015in). The preparation of one hide could take up to eight hours, and a body covering would require four to five hides. In the harsh climate of the Pleistocene, new clothing would be needed at least once a year. The economic parallels of this are astonishing: today, many could easily get by on two weeks' salary as an annual clothing budget.

These bodies show us that movement was written in the bones of our prehistoric forbears. Like a fingerprint from the past, the bones document their use at levels that seem incomprehensible to an Anthropocene human.

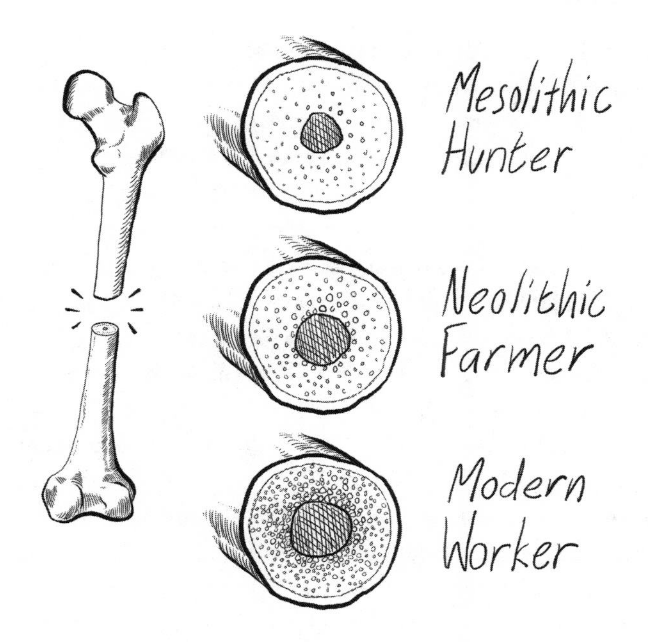

Research shows that as we progress technologically through each lifestyle revolution, our bones get increasingly thinner and more porous.

ACCELERATING CHANGE

As the climate dried out several million years ago, converting the forests of central Africa to grasslands, humans emerged with bodies ideally suited to this new environment. There, they thrived and roamed. In constant motion, they moved every day and migrated slowly across the globe. Through constant variation and mutation, new species families emerged, converged and died out. Life was hard. Mean life expectancy would have been short. These earlier versions of ourselves may not have struggled with the hundreds of diseases and debilitating conditions that we have, but they lived precarious lives. An impacted tooth could kill you, as could a simple fall or the slip of a hand during tool manufacture. Remaining wary of the kinds of nostalgia that M R James so adeptly warned about, in which the brutality of the past was laid bare, there is also the fact that the bodies of these paleo humans did not suffer the levels of morbidity that are common in modern humans.

Only *Homo sapiens sapiens* made it to the Holocene; although some other species almost made it, they are now all gone. But they have left behind ghosts of themselves in our DNA. The Denisovans are to be found camping out in the cells of indigenous Australians and Papua New Guineans. The Neanderthals have found a quiet corner for themselves in the genes of most East Asians and Western Europeans (making up about 2–4 percent of their DNA).

Humans are now the single greatest factor affecting evolutionary change. We are forcing the evolution of our environment and making it happen at a pace that neither we, nor other animals and plants, can respond to effectively. As a result, the world is becoming a genetically simpler place.

In the Holocene, species extinction was a natural phenomenon that occurred at about the rate of one to five species every year. In the Anthropocene, that figure lurches between 1,000 and 10,000: the equivalent of dozens every day.[4]^

The two body parts that have been the focus of this chapter, like so many others that followed, underwent substantial change during the approaching millennia. This seems fine; life was about change. The problem is that the code for the morphology and function of these body parts remained unaltered. Our feet are longer, thinner and flatter than those of hunter-gathering tribespeople, but that is not because our DNA has changed.

While there are factors that can alter our DNA during our lifetime, such as exposure to radiation, or in some cases to alcohol (which recent research suggests can damage DNA in stem cells^), some believe that ageing is the accumulation of DNA damage that is left unrepaired. Despite this, the genetic code activated in a mother's womb at the moment of conception is the same as it ever was. And when the feet of a modern child first come into daylight, they still resemble those of children born on the savannahs of prehistory. But within that genetic code are lines and sequences that respond differently when they meet certain external physical stimuli.

As innovation and change began to spike in the Holocene (which constituted less than 0.5 percent of our history), we have repeatedly mistaken what is comfortable with what is best for us, what is ideal for that which is

---

4 According to the World Wildlife Fund, freshwater ecosystems are particularly endangered, which is important because of the 25,000 species of fish on the planet, 40 percent are freshwater. "In the United States it is estimated that 54 percent of original wetlands have been lost, 87 percent to agricultural development. In France, 67 percent of wetlands have been lost in the period 1900 to 1993." Evolutionary change should be something that happens relatively slowly, but the pace of innovation and change is being squeezed like toothpaste from a tube in the rush of a Monday morning. Like our environment, our bodies are also struggling to keep up.^

quickest or easiest, and what feels good for what is good for us. And as the transformations, inventions, comforts and technological solutions pile up around us, it seems that a digger's claws have opened above our heads and showered them down on us – our bodies are struggling to keep pace with the changes we have made and are continuing to make.

WINDING BACK

Move along, folks. Nothing to see here…

Despite the fact that the Palaeolithic human underwent significant hardship, many of the pathologies that humans are riddled with today were yet to emerge.

In the upcoming revolutions (Agricultural, Industrial and the sedentary ones), there were fundamental changes to our way of life that introduced or significantly accelerated existing pathologies. Although there were advances during this period (tool manufacture and use, collaborative hunting, the emergence of language and art) these did not bring with them immediate and revolutionary changes that were inscribed on the bodies of early *Homo*.

The savannah was a place in which human DNA met the environment equably. "Perfect" would be too strong a word, but the human body was able to flourish in that environment. Humans did not, to anyone's knowledge, pick up habits or develop practices that have proved to be harmful in the millions of years that followed – if they had, could they have lasted so long there?

As we head into the upcoming anatomical revolutions, and things begin to go awry with the human body, these sections will look at how revolutionary changes to the ways in which humans have lived have become habitual and normal to us today, having a profound impact on our morphology, movement and health.

You can never go home again, and as L P Hartley told us, "the past is a foreign country"^ where they most certainly do things differently; we cannot go back to the Pleistocene. And if we could, who among us might want to? Instead, the remaining Winding Back sections will be about finding ways of integrating practices or modes of behaviour that served the human body well in the past.

They will provide a snapshot of what changed during these revolutions, and look at how the human body rewards certain behaviours and activities that can easily be imitated and incorporated into even the busiest modern lives.

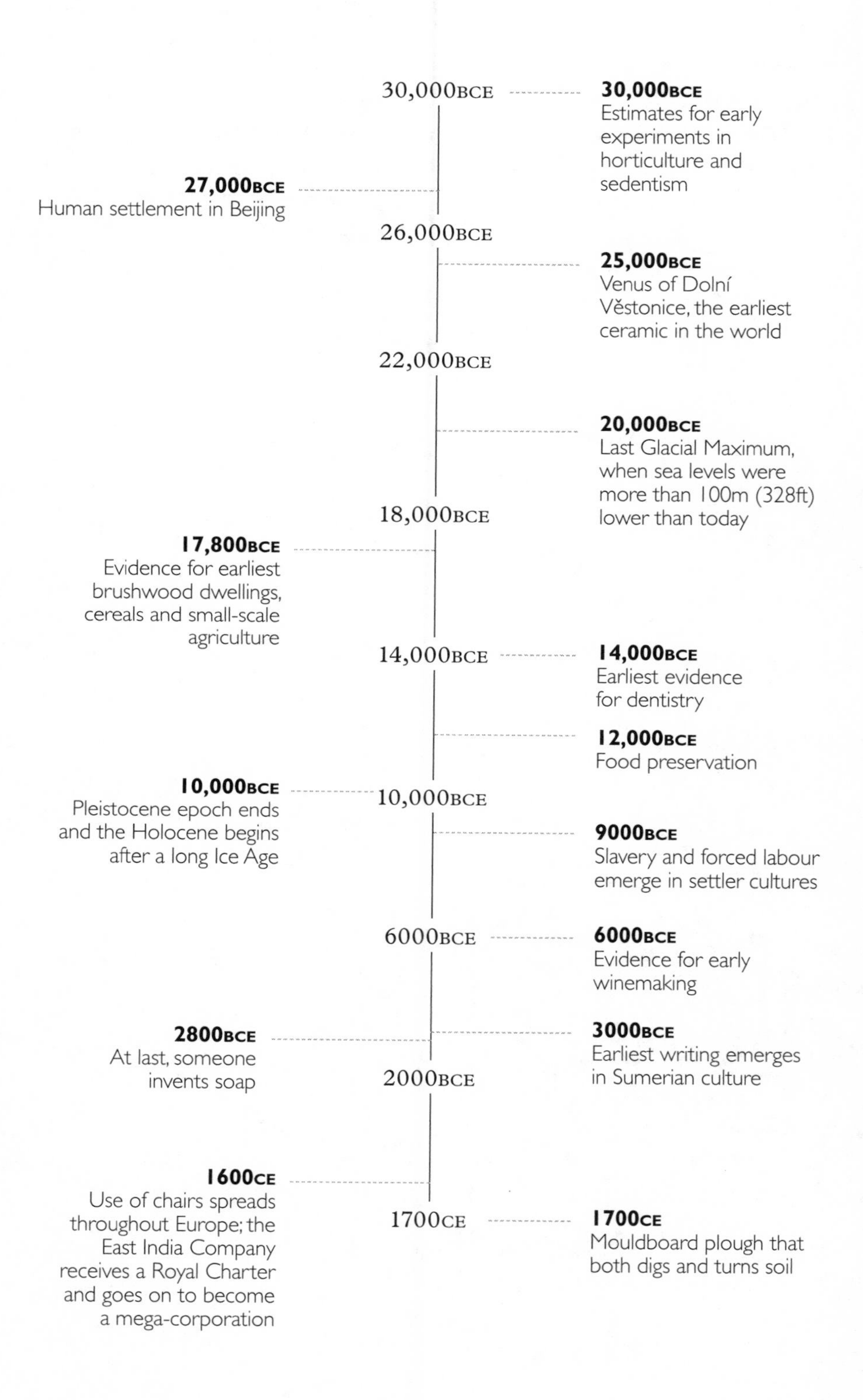
30,000BCE
30,000BCE
Estimates for early experiments in horticulture and sedentism
27,000BCE
Human settlement in Beijing
26,000BCE
25,000BCE
Venus of Dolní Věstonice, the earliest ceramic in the world
22,000BCE
20,000BCE
Last Glacial Maximum, when sea levels were more than 100m (328ft) lower than today
18,000BCE
17,800BCE
Evidence for earliest brushwood dwellings, cereals and small-scale agriculture
14,000BCE
14,000BCE
Earliest evidence for dentistry
12,000BCE
Food preservation
10,000BCE
Pleistocene epoch ends and the Holocene begins after a long Ice Age
10,000BCE
9000BCE
Slavery and forced labour emerge in settler cultures
6000BCE
6000BCE
Evidence for early winemaking
2800BCE
At last, someone invents soap
3000BCE
Earliest writing emerges in Sumerian culture
2000BCE
1600CE
Use of chairs spreads throughout Europe; the East India Company receives a Royal Charter and goes on to become a mega-corporation
1700CE
1700CE
Mouldboard plough that both digs and turns soil

Part II

30,000BCE–1700CE

SEEDS, SETTLEMENTS & CITIES

For there was never yet philosopher that could endure the toothache patiently.

Leonato to Antonio in Shakespeare's *Much Ado About Nothing*

As the Pleistocene was drawing to a close, preparing to shift into the more equable Holocene, there seems to be evidence that the intellectual capabilities of humans began to sharpen. But, if there ever was a Cognitive Revolution, it probably happened over tens of thousands of years. The supposed time lag between the arrival of the modern body and that of the modern mind (hundreds of thousands of years later) has left anthropologists scratching their heads.

Recently, Sally McBrearty and Allison Brooks presented evidence suggesting that we need an even greater timespan from 280,000–50,000 years ago to explain the emergence of behavioural modernity. In "The Revolution that Wasn't: a New Interpretation of the Origin of Modern Human Behavior"^ they explained that a number of technologies and behaviours associated with the Cognitive Revolution emerged in Africa much earlier and that the widely held view that "the earliest modern Africans were behaviourally primitive… stems from a profound Eurocentric bias… These features include blade and microlithic technology, bone tools, increased geographic range, specialized hunting, the use of aquatic resources, long distance trade, systematic processing and use of pigment, and art and decoration. These items do not occur suddenly together as predicted by the 'human revolution' model, but at sites that are widely separated in space and time."

By the arrival of Holocene, then, humans had already acquired the skills and desire for personal adornment and jewellery. Evidence for religious practices and rituals emerges, too, with rising belief in a soul, afterlife, shamanism and mediumship. These coincide with the appearance of Palaeolithic cave art depicting animals such as deer, horses, rhinos and cows. There was little narcissism in early art; depictions of humans were rare, though hands were outlined and other body parts such as genitals were scratched into walls. When humans did appear they were often much more abstractly represented than the animals. This artistry transforms into anthropomorphism, sculpture and early symbolism, which were all present about 30,000 years ago.

It would be wrong to ascribe this inventiveness only to the species *Homo sapiens* as the age and location of many cave paintings makes it more than likely that they were made by Neanderthals exercising their symbolic and creative capacities (*sapiens* seem oddly unique as a species in their unwillingness to endow other members of the genus, or indeed other species, with skills they like to claim are unique to themselves; art is one such example, but communication, laughter, deception, a sense of self and tool use are similar examples found in other species).

For a couple of million years, the kinds of work done by the human body had remained relatively consistent. Once early humans migrated, the variety of work that they undertook changed, but it would have remained relatively stable within a group or tribe. Once symbolic art began, it presents a clear example of a shift that divided labour and skill between individuals. Sculpture using basic tools required great expertise, not only to create a symbolic object, but in acquiring the skills necessary to create an object. To allow one person to work on an ivory carving, extra work had to be done outside daily by others to maintain and support that work, possibly for many hundreds of hours in total.

With the rise in cognition came a proliferation in the kinds of work done by the human body. One way of thinking about the variety of labour that existed through the majority of our history might be as an ecology. This is a 19th-century word for a branch of biology that examined the relationships between organisms and their environment. The term quickly widened to include the more sociological aspects of these relationships and dependencies, looking at the connections between people, either as individuals or as groups, and their environment. And it widened again to include the systemic relationships that emerged in areas of human settlement.

Our relationship with our environment is mostly founded upon work. The organism that performs the work must be adapted to complete it; if it is not, then some adaptation to its soft and hard tissues will take place. If the work goes beyond the physiologically adaptive capacity of the organism, it will struggle in the environment or die.

The ecology of labour that predominated for more than 99.5 percent of human history (roughly the length of time it took to reach the Agricultural Revolution) was one that included a fair amount of variety. Like any ecosystem, balance can be found in such variety. And although these early humans were plagued by health problems which could be easily cured by modern medicine, their functional fitness meant that they were not utterly overwhelmed by biomechanical and metabolic pathologies. Their lives were harder and their stomachs emptier, but when all the work was added up (including domestic maintenance, food preparation and everything else), hunter-gatherers would have worked in total perhaps 30–40 hours a week, but probably less than this.

Compared to Western societies, where most people work 40–50 hours a week, the relatively short working week of *Homo erectus* or *heidelbergensis* does not seem so disparate. Today, many of us are martyrs to email and gratefully accept the heroism that comes from working a little harder than our predecessors. But once we factor in *all* work (that is all domestic maintenance, financial management, emails at the weekend, shopping, cooking, cleaning – all those things we might think of as leisure) the computer calculating the Western workload would go "poof" – springs and smoke boinging from it as if Wile E Coyote had rigged it up and, grinning, plunged the detonator.

The trouble began when humans stopped walking. When did a band of tribespeople suddenly decide "enough"? And what happened to their bodies when they did?

This is the time during which we see the beginnings of the modern body. A moment arrived perhaps 30–25,000 years ago, when someone somewhere realized that here at last was a place with access to water and an abundance of resources such as food and simple construction materials, and that there was no reason to move on. If this was a "moment" it was the single most important one in our history. Even so, nobody knows whether agriculture was an outcome of sedentism, or if sedentism emerged as a necessary by-product of agriculture.

In this new environment, human range of movement began to diminish slightly. People began to walk a little less, climb a little less, hunt *a lot* less; some already wore shoes. They were growing cereals, which means that they worked machines such as grindstones, experimented with irrigation, storage and preservation. They were using tools more regularly, sleeping in shelters, keeping house, domesticating and rearing animals – at first to eat, but later for skins, milk and to put to work. More mouths had to be fed. These subtle changes in behaviour began as a few acorns and grew into forests.

Once humans stopped roaming, their bodies shrank and began to change. As if marking disapproval of their new-found adoration of a carbohydrate diet, average height also dipped quite dramatically during this period. The bones of settlers, although still much stronger than ours today, began to thin. Not only did the new food they began to eat have an impact on the state of their teeth and their organization, it frequently changed the number of them in their mouths. There was also the fact that their mouths suddenly seemed more crowded, despite their teeth shrinking, and that their teeth rarely seemed to fit into the space available to them. Why was this?

For a couple of million years humans had existed outdoors, but once shelters, settlements and cities began to rise, more time was spent doing work while shaded from the sun.

Because we are in the realms of prehistory, before writing culture began recording historic events, we can only infer, initially at least, from archaeology and the fossil record. Writing arrived long, long after sedentism. By the time people had begun pressing letters into clay, they were already ensconced in cities surrounded by high walls, and some of these had many tens of thousands of inhabitants.

In evolutionary terms the Agricultural Revolution was a period of sudden change and it is interesting that even though those early settlers seem so distant from us, their bodies were slowly reshaping to look more like our own, changing their frowns and their smiles into ours.

SETTLING IN

In a region all too familiar with intense periods of drought, Israel in the 1980s suffered one of its worst. It is often the case that exceptional circumstances give rise to exceptional discoveries in archaeology. During one of the most serious dry spells the region had ever known, in 1989 the shore of the Sea of Galilee in Israel exposed a prehistoric settlement that had been preserved there for many thousands of years.

Ohalo is the Pompeii of the Agricultural Revolution. The archaeological site (called Ohalo II), dated at about 20,000BCE, is an ancient waterside settlement that was preserved when it was submerged, probably during exceptionally heavy rainfall^, or in one of the many floods that form our myths of origination – from Noah's flood in the Bible to the ancient Babylonian epics of *Gilgamesh* and *Atrahasis*. The water level rose, covering the site in sand and silt. When it was excavated it was an anthropologist's gold mine.

They discovered numerous domed brush huts constructed from oak, willow and tamarisk. This shape would later give rise to the massive super-dome structures that appeared in the cathedrals of most major cities throughout the world. In the Ohalo huts were floor coverings and bed mattresses (*mattresses!*) made from bundles of cereal stalks.

There was also evidence that residents had eaten fruit, wild vegetables and the food that is the mainstay of the Anthropocene human diet: cereals. Anthropologists discovered gathered grains of barley and wheat in their thousands. For settlements discovered from this period, this is exceptionally rare. Remains showed that people also hunted large animals (wild deer and gazelle) and were capable fishers.

The disinterred remains of a 40-year-old male (*see* opposite) had a right hand and arm which were much more developed than his left, suggesting that he was a spear-throwing hunter or archer – or more likely did heavy manual work. He had been seriously injured and had probably been left disabled with an open wound for several months. His remains, though, suggest that he had been taken care of, and when he died he had been thoughtfully buried, arms crossed over his chest, legs curved in parallel behind him, revealing a strong sense of social cohesion among a group that cared for its elders and its sick.

Agriculture brought with it the benefits of social care. Once the immediate need for food was met by agriculture and its associated technologies, humans began to look for other fields to plough.

Ohalo II shows us agriculture in its heyday. There is some indication of cultivation on a small scale with evidence of a mix of crops (rather than large fields of them), and flint sickles have been discovered that show only moderate wear. And the more humans grew used to these new food sources, the more their – and our – appearance slowly began to merge.

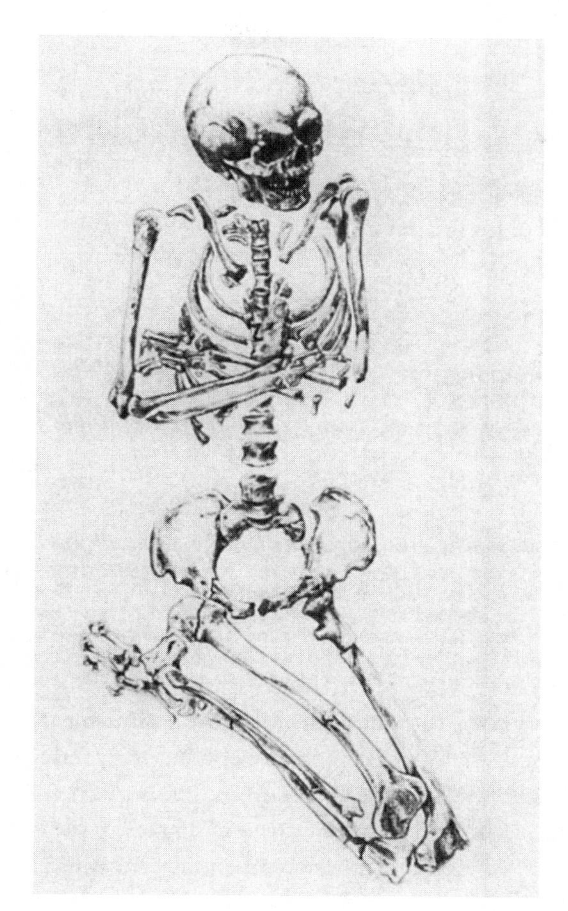

Sketch of the buried male skeleton discovered at Ohalo II.

CHAPTER 3

PLANTING SEEDS, PROCESSING FOOD & LIFE UNDER COVER

> ... the lust for comfort, that stealthy thing that enters the house a guest, and then becomes a host, and then a master.
>
> Kahlil Gibran, *The Prophet*, Wisehouse Classics (2015)

For tens of thousands of years, humans collected wild grains to eat them. The kind of archaeological evidence that would tell us exactly when this happened is unlikely to be discovered. The conditions for the preservation of plant remains have even tougher criteria and prerequisites than human fossils. But there is evidence that humans were experimenting with horticulture 30–20,000 years ago, although full-blown agriculture took a little longer to emerge. Evidence shows that grinding stones were used to refine and process grains to facilitate consumption. In fact, there is widespread evidence that certain kinds of food processing were well under way. Archaeologists Ainit Snir and Dani Nadel explore the evidence for the processing of wild wheat and barley as much as 23,000 years ago in "The Origin of Cultivation and Proto-Weeds, Long Before Neolithic Farming".^

There is evidence for other kinds of food processing long before this. Cooking meat and vegetables, as well as other kinds of food preparation, were widespread. Among the techniques used were a tin foil method in which foods were cooked over hot ashes wrapped in leaves (especially good for delicate foods such as fish). Birch and willow branches could be shaped into a kind of grill to place over a low fire or hot ashes. Wide, plate-like hot stones forming a bridge over a fire functioned like a baking tray or frying pan. And there was also the spit – a mainstay of many a barbecue today.^

Although these changes were slow to gather pace, the impact they had on our species was profound. It is currently estimated that Americans between them eat many billions of hamburgers each year.

The stats tell us that nearly everyone on the planet loves a burger. (I'm a not-very-good vegetarian, and I miss them.) Their popularity tells us something about what we like, set against what our bodies need. The average American eats approximately 28kg (61lb) of beef each year. Imagine chewing your way through half your body weight of steak. Imagine the pain in your jaws and temples. Imagine how long it would take to tear the meat with your teeth into chewable chunks and to grind and work it with your molars, ten, thirty, fifty or sixty times to reduce it to a swallowable and digestible state. My guess is that this would lead to the sorts of stress fractures in mandibles that new runners sometimes suffer when they overload their frame.

Most of our meat is consumed in burger form, which means that not only has the meat been cooked, but the beef has already been ground up. These things are not inherently bad. Some research has even postulated that it is the cause of the speculative bump in human IQ during that so-called Cognitive Revolution:^ it derives from human access to meat sources and our ability to absorb their nutrients through cooking and food processing. But today we even have our chewing done for us at a mega-factory probably thousands of miles away from where we live and eat. This is the most flagrant kind of outsourcing imaginable. And of course, it's not just burgers that are processed. The majority of the Western diet consists of foods like these, all are astonishingly easy to digest, so the body can whip all the calories and sugars out of them with the greatest ease.

All this is changing the line of our jaws, the alignment of our teeth, the shape of our faces, how we look. Our food is contributing to the creation of the Anthropocene face.

TERRIBLE TEETH – AN ARCHAEOLOGY OF PAIN

To begin the story of how our faces have changed we need to look at our very own food processors: our teeth. Talking about our teeth is as close as we can get to recounting the story of our entire evolution. In them, we can see links to our ancestors, to other mammals – and their wear and tear tells the story of both culture and environment.

Evolution was still, 20,000 years ago, doing its ever-changing work in subtly altering human looks, with the alleles of DNA doing their constant shifting dance of variation and mutation across and through the species. To give a brief idea of the amount of variation that is possible within a human species, if you take a male and a female, during reproduction each will contribute a random selection of 23 of their own chromosomes to their child. The total number of variations from just these two parents' coupling comes to a total 2^{46}. That's more than 70 trillion possible outcomes – many more times the number of all the humans who have ever lived. Once you factor in the crossing over, or partial recombination, of DNA (in which parental genes randomly swap chromosomes), the possibilities for variation become infinite.

This means that all our faces are different; our eyebrows, our noses, our lips, our irises – everything. But in many cases our bodies rely on stimulation or sustenance from the environment of their habitat to develop properly. The body expects a certain place with specific resources and one that demands and encourages behaviours that will reward it, and if that environment is not there, changes and mutations occur. This is what happened to the first agriculturalists, and just as with other stories about changes in human height and longevity, their new habitats changed them. Changes to teeth and jaws come sharply into focus during this period, but if we are to understand their development, we need to look a little at their history, too. What sort of teeth did humans have when they first ground up grain? Where had they come from?

Among primates, our teeth are unusual for their comparative size and position. They are thrust quite far back in our heads (which helps us to maintain a good centre of gravity while standing or moving on two legs – particularly running). Our jaws are relatively small, and so too are the teeth in them.

In earlier hominins, such as Lucy (*Australopithecus afarensis*) and *Paranthropus boisei,* teeth were much larger than ours today. This was not the case for all hominins; the diminutive *Homo floresiensis*, with a skull only a fraction of the size of a human's, had smaller teeth than those of their *sapiens* cousins. The east African *Paranthropus boisei* lived from 1.2–2.3 million years ago and were diminutive in height; males were about 137cm (54in) tall, so they were basically all jaw, with teeth nearly double the size of ours. Their

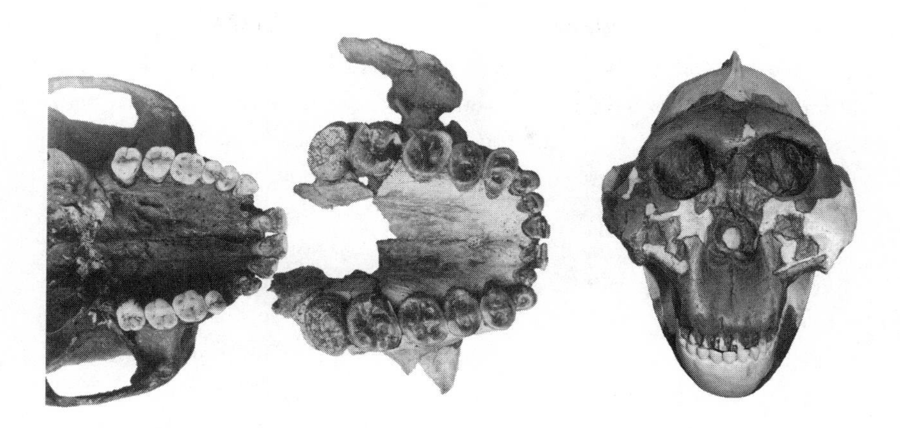

A *Paranthropus boisei* jaw (centre) compared with a modern human (left) and the skull of *Paranthropus boisei* (right). The *Paranthropus* jaw and teeth are much larger, especially given that they were up to 60cm (2ft) shorter than humans.

skull structure included a sagittal crest, a clearly-defined ridge that runs along the middle of the skull (like a bony Mohican haircut) indicating that exceptionally strong jaw muscles were attached there, making it possible for this species to survive in places where the only foods available were exceptionally tough and fibrous.

For some reason, the large chewing teeth, with comparatively thick enamel and large plate-like roots, had shrunk to a fraction of their previous size by the time they made it into the mouths of taller *Homo sapiens*. The molars in earlier species had about ten times the volume of the front teeth. Humans today have teeth only two or three times larger, and the lower jaw is so small that it can rarely house a third molar (a wisdom tooth).

What is overwhelmingly significant in the fossil record of the teeth of early humans is their condition. Our teeth in the Anthropocene are not like those in the jaws of our ancestors. Malocclusions (or dental crowding), with misaligned and crooked teeth, are now so common as to be normal (with serious cases in as many as 20 percent of the population in high-income nations). This is worse for some more than others and they are required to wear braces or retainers to straighten their teeth for several months. Sometimes this is cosmetic, but mostly the problems show that something has gone awry during tooth and jaw development.

Earlier hominins, whose teeth were much larger than our own, show almost no evidence of needing a brace (there is a *Homo floresiensis* skull from about 18,000 years ago that has a rotated fourth premolar – which happens in modern humans where the tooth turns in the gums to accommodate growth around it – but these are quite rare). In hominin species something went wrong as our jaws and mouths shrank in size. In pre-agricultural times, *Homo sapiens'* dental crowding and crookedness is not very easy to find. After that period, with agriculture in full sway, it is as if our teeth no longer fitted into our head.

Dental crowding seems to have become more common when humans started growing and rearing rather than hunting and gathering.

Cavities reveal a similar pattern. The rates at which dental cavities appear in the fossil record of our pre-agricultural ancestors are very low. They are approximately: *Homo erectus* (2 million years ago) 4.6 percent; *Paranthropus robustus* (1.2–1.8 million years ago) 2.6 percent, and *Homo naledi* (236–350,000 years ago) 1.36 percent.^

Humans today understandably show huge variation. There is no robust global data, but in the UK 7.31 percent of adults have tooth decay and 74 percent of adults have had a tooth pulled out. Only one in ten adults has excellent oral health. But compare these numbers with a group such as the Turkana people of north-west Kenya who clean their teeth with brushes made from the esekon tree, where cavities are practically unheard of. Existing as the Turkana do on an unprocessed diet in which the only sweetener available is a little honey, it seems that the numbers clearly signal something unique to post-agricultural humans.

EVOLUTIONARY WISDOM – PROBLEM TEETH BEFORE AGRICULTURE

Although many modern tooth problems emerged when we turned more toward cereals, beans and crops, the earliest evidence of cavities in hominins is about 1.5 million years old. It is difficult to make solid ancestral comparisons between *Homo sapiens* and other hominins because of the much smaller samples we have for older species. There are specimens from Swartkrans, a cave in South Africa, and a *Homo heidelbergensis* (about 500,000 years old) from Kabwe in Zambia which have bad cavities. In terms of underbite and overbite, a small overbite is normal but there aren't many early jaws to check this against, as it is rare to find the upper (maxilla) and lower (mandible) jaws which are reliably known to belong to the same individual. For many species, there might be a single jaw, or perhaps a sample of ten to fifteen of them, but these few then represent thousands of people over thousands of years and rarely give an accurate snapshot of what a group might have been like in terms of dental health.

On the one hand, teeth are a gift to archaeologists and evolutionary anthropologists because they are so tough. Sometimes all we have of a fossil is a tooth or two – the teeth are often complete. But a single tooth tells us a lot. Teeth are the analogue of our bodies and their DNA. Just as the environment changes us, from the moment a tooth appears in the mouth, it is subject to damage, chips, scratches, decay and wear. It becomes a palimpsest, a surface on which stories and histories are etched before they are wiped clean to be reused. While we use our teeth to chew things, we are not always very responsible owners; has anyone opened beer bottles with theirs? Many apes use their teeth as tools to crack nuts – but teeth, unlike bones, never remodel during life.

Many of the teeth in the fossil record are damaged. The other symptoms you might find are chips caused by a diet which may have been rich in gritty foodstuffs such as tubers; and there's also hypoplasia – an interruption in the formation of the enamel – that caused pockmarks and ridges to develop. The lines might have been caused by a serious illness or malnutrition but could also have come from an annual cycle of resource scarcity.

One outstanding question from looking at practically any other member of the hominin family concerns the relative size of our jaws. Should they be bigger than they are?

The simple answer to this is: yes. The reason they are not larger is the extent to which food is now processed away from our mouths. And because the bone can change throughout your life, chewing really hard from when you are four until you are thirty or forty years old will promote bone remodelling to support your jaws in exactly the same way as a runner's hips, knees and ankles grow stronger from their impact loading.

The opposite scenario emerging during this period in which our diet changed, and which is nearly inevitable in modern life, would be that the jaw becomes less robust, with thinner bone surrounding the teeth.

A 2014 article, "Age Changes of Jaws and Soft Tissue Profile"^ explained that during human adolescence the face widens as the growth of the lower jaw exceeds the upper. The length of our femurs (thighs) changes dramatically during adolescence, when we grow to our full adult height, but there is only a brief time when this growth can happen. With the right diet the body can complete its growth spurt, but after that we cannot start eating a healthier diet in our mid-twenties and hope that our femurs will, at last, lengthen. The jaw is probably same, its remodelling is as diet dependent as our height. Research would suggest this is so.

Harvard-based evolutionary biologist Daniel Lieberman looked at the craniofacial growth patterns of hyraxes (*Procavia capensis*) and found that processing raw food required as much as twice the force from the jaw than eating cooked food. In the experiment, hyraxes on a cooked food diet showed significantly less growth (as much as 10 percent) in the lower and rear parts of the jaw. "The results support the hypothesis that food processing techniques have led to decreased facial growth in the mandibular [lower jaw] and maxillary [upper jaw] arches in recent human populations" according to Lieberman.^

Noreen von Cramon-Taubadel at the University of Buffalo New York, has worked on precisely this issue in humans. Her 2011 study, "Global human mandibular variation reflects differences in agricultural and hunter-gatherer subsistence strategies"^ set out to provide some sense of how variations in diet affect the shape of the jaw. She explains that the substantial variation in the chewing behaviour of hunter-gatherers and modern populations is thought to be one of the principal forces affecting the shape and size of the lower jaw, and she looked at whether this hypothesis played out globally. She found that the lower jaw (unlike the rest of the skull), "significantly reflects" the diet and means of subsistence, rather than the inherited genetic patterns of the subjects.

Hunter-gathering peoples had jaws that were both narrower and longer than those of populations that subsisted on processed and cereal diets. Cramon-Taubadel's results support the idea of a decrease in the stimulation of intense chewing found in the hunter-gatherer diet compared to today's pappy modern diet: "the developmental argument also explains why there is often a mismatch between the size of the lower face and the dentition, which, in turn, leads to increased prevalence of dental crowding and malocclusions in modern postindustrial populations."

Another study of the fossil record analysed 292 specimens, geographically and temporally spread and found "harmony" in the jaws of hunter-gatherers.^

The modern mismatch disease of an underbite or overbite seems to have emerged during the change to a diet based on legumes and cereals. The study's authors found that while hunter-gatherers' teeth comfortably fitted their jaws, with a correlation between mandibular and dental distances, "in the case of semi-sedentary hunter-gatherers and farming groups, no such correlation was found, suggesting that the incongruity between dental and mandibular form began with the shift towards sedentism and agricultural subsistence practices."

Our jaws have not returned to the size they once were; the kind of raw strength they once had is not necessary to eat a diet that consists mostly of soft foods. They are continuing to shrink because of our lifestyle, and so it seems are our teeth.

Our teeth are *much* smaller than they used to be, especially if one compares a modern human skull with that of the vegetarian *Paranthropus boisei*, with their massive sagittal crest, big cheekbones, outsize teeth, thick enamel

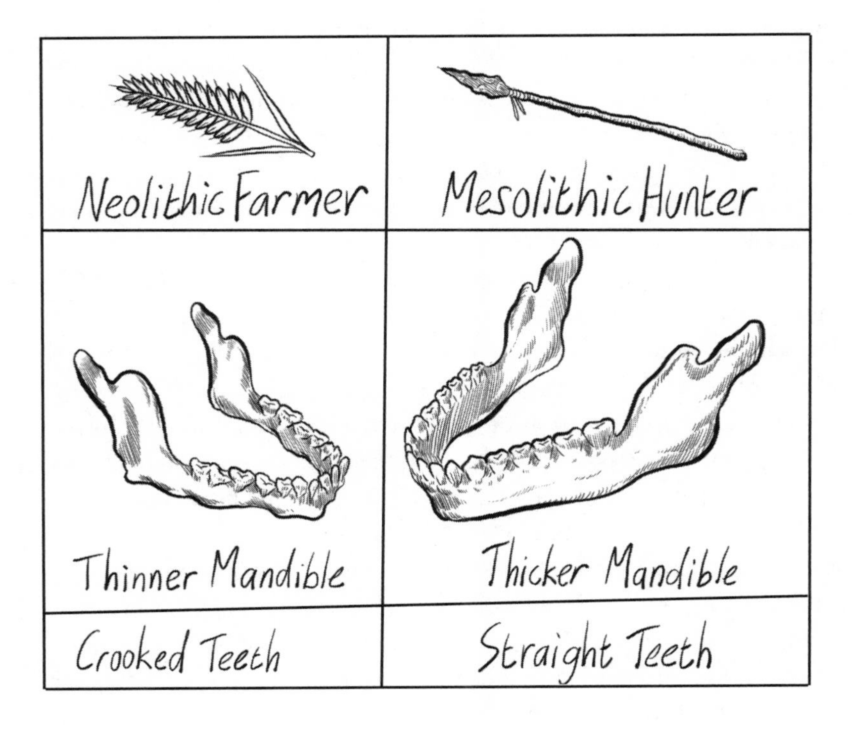

and more complex roots. Everything has been selected for seriously heavy chewing. But there are different ways in which human teeth have become smaller: our molars have become smaller from the back. In the past, the third molar (wisdom tooth) tended to be the biggest tooth in the jaw. But in modern humans the first molar is usually the king of the pack, reducing the risk of impaction at the back of the mouth.

Teeth are made up of enamel and dentine, so if there was favourable selection for smaller teeth, that could happen in a number of ways. In humans it seems that there has been an overall reduction in the volume of the dentine, leaving the more protective enamel in place.

We are living in a time when dentistry is widely available and the idea of tooth impaction seems only uncomfortable, with the very worst outcome being extraction and a course of antibiotics, but for our ancestors it was a potentially fatal design flaw.

Tooth infections could lead to pus leaking into the jaw and it rotting. In such a case, pus would also leak into the mouth. There are good examples of this in the fossil record, such as a Neanderthal (who had only about a one percent risk of cavities) from Sima de los Huesos in Spain who had an impacted tooth of such severity that it had deformed the bone. This would have made it significantly less likely that the subject would have reproduced, not only because a mouth that had pus leaking into it was not a particularly attractive prospect, but also because the condition, which could lead to serious infection, was life-threatening. Natural selection for smaller teeth lowers the risk of such impactions.

Evolution is still doing its work with our wisdom teeth. Many people reading this will not have all of them. Some will have had them removed; others, like myself, will have had only one set of wisdom teeth come through, with the other set staying out of sight beneath the gum line. But many don't have them at all. Their DNA is programmed to save them from the horror, peculiar pain, embarrassment – but mainly the risk – of impaction. It's one of only a few examples of evolution at work in our bodies today.

There are fossils in China that date at their latest from about 300,000 years ago which have no wisdom teeth.^ During the last few hundred years, healthcare will have interfered with the work of evolution, but before then, having no wisdom teeth would have been a selection that favoured survival and procreation. In the space of just 300–400,000 years, this genetic trait has spread to an estimated 2.5 billion people who have at least one missing wisdom tooth, with many having none at all.

Evolution may be slow at solving such developmental problems, but it is trying. In the 10,000-or-so years since our diet changed so radically, it is attempting to correct the lack of space in our jaws created by the foods we now eat.

One of the problems with the "paleo" diet is that no one really knows what early hominins ate. We can make intelligent guesses inferred from the animals they hunted, but precisely which fruits, berries, tubers and insects were part of their diet we cannot be sure. A study called "The plant component of an Acheulian diet at Gesher Benot Ya'aqov, Israel"^ was able to show that, at least in the case of the Levantine corridor 780,000 years ago, humans apparently ate and cooked anything they could get their hands on. "The food plant remains were part of a diet that also included aquatic and terrestrial fauna", providing evidence for "a varied plant diet, staple plant foods, environmental knowledge, seasonality, and the use of fire in food processing." Among the meats consumed were elephants, hippos and rats. "Paleo" diet, anyone?

From excavations at the sites of early agricultural settlements we have a much better idea of what agriculturalists consumed, because just like us, they left lots of waste behind. What we can be sure of is that the consumption of carbohydrates quickly escalated. In both cases, humans are shown to be seasonal opportunists and organisms highly adaptable to their environment.

All foods can be broken down into seven essential components: water, fibre, vitamins and minerals, and the three macronutrients – protein, fat, and carbohydrate. Carbohydrates, including starches and sugars are built from simple sugars such as glucose, fructose and galactose. These monosaccharides are so easy for the body to use they are not even digested, just absorbed and broken down inside cells. When two simple sugars are joined together by chemical bonds such as carbon, hydrogen or oxygen, they become disaccharides (sugar or lactose). Starch is an example of a polysaccharide because it consists of several glucose units joined together by glycosidic bonds. The growth and harvesting of foods comparably higher in carbohydrates meant that as the human diet began to shift into this new phase in which the balance of macronutrients changed, it brought with it a more sustainable energy source, and a satiation that allowed these early farmers to concern themselves with other questions, tasks and aspects of life.

Horticulture, and the reaping of a small harvest, meant that more time could be spent creating better and hardier shelters. The crops from a garden could also feed more people. Starch is produced by most green plants as an energy store, and eaten by humans it becomes a similar source of energy. The problem is that, just like the acidic environments that quickly destroy fossil remains in jungles, a similar environment is encouraged in the human mouth when eating sugars of any kind.

The tooth surface, which is non-renewable and finite, is exposed to certain bacteria in the mouth that are fond of fermentable carbohydrates. As bits of food adhere to the teeth, bacteria that live in the mouth go to work metabolizing it, which results in an acidic by-product called plaque. Plaque collects on the enamel of our teeth, particularly around the gums. For those whose diet is low in sugars this presents a lower risk because the environment provides less opportunity for the bacteria to thrive. But when humans eat an agricultural diet in which sugars become a macronutrient, there are widespread changes to the ecology of the mouth.

As we have evolved as a species, we have also influenced the evolution of the bacteria that live in our mouths. A genetic study of the bacteria *Streptococcus mutans* showed that about 10,000 years ago several of its genes were positively selected – and these were precisely the genes involved in "sugar metabolism and acid tolerance". Another 73 of its unique genes could be "associated with metabolic processes that could have contributed to the successful adaptation of *Streptococcus mutans* to its new niche, the human mouth, and with the dietary changes that accompanied the origin of agriculture."^

The bacteria that created our plaque *loved* our new diet, so it evolved to rot our teeth all the more efficiently.

Even though our teeth are without doubt the toughest and hardiest things in our bodies (the entire Denisovan's fossil library consists of only three teeth), when something goes wrong with them, it affects the whole body. Nothing is so all-encompassing as toothache, and although the teeth of hunter-gatherers were much healthier than those of early farmers (or, of course, our own), there is evidence that they struggled. Dentistry, it seems, has a surprisingly long history.

In June 2017, *The Bulletin of the International Association for Paleodontology* published a study of some old remains – in both senses of the word. They were Neanderthal but were excavated more than a century ago from a site in Croatia. The team of researchers used a variety of techniques to analyze the marks and scratches on several teeth from a particular specimen. What they discovered throws new light on our understanding, not necessarily of prehistoric dentistry, but at the very least, of the ingenuity of earlier humans in having a go at it themselves.^

Several of the teeth showed wear of a kind that could only have been caused by a sort of toothpick. The scratches and grooves suggest that the owner of these teeth was probably suffering with periodontal disease, but because the teeth were not attached to the mandible, this could not be confirmed. The dents and chips look as though the individual was angling to reach a twisted premolar to alleviate some discomfort. What is unusual about this evidence is that it is so much earlier than any other – this act of auto-dentistry dates to approximately 130,000 years ago.

Other evidence suggests that pre-agricultural peoples used chewing twigs, feathers, shards of bones and even porcupine quills to remove food debris from their teeth.

There is an early example of dental cleaning with a little drilling too from around the time our diet was changing. A team of researchers led by Stefano Benazzi at the University of Bologna examined the teeth of the remains of a young man.^ His teeth dated him to between 13,820 and 14,160BCE. The researchers suggested that the tooth they had studied had several cavities. When they looked a little closer at the largest of these, they discovered peculiar marks inside, suggesting that the infected tissue had been picked away with a small, fine, stone tool – it had been scooped and dug out as if drilled. This was a specific tool, produced for a specific purpose, and makes the procedure look a lot like pre-Neolithic dental medicine. This is the earliest known example of tooth picking to remove tooth decay, rather than just food debris.

Pierre Fauchard was an 18th-century French surgeon who was said to have invented the filling, but a discovery in 2012 by an Italian team found a Stone Age tooth in Slovenia that had been filled with beeswax, suggesting that Fauchard might have missed his moment by about 6,500 years. The team reported "the earliest known direct evidence of therapeutic-palliative dental filling."^

During the Agricultural Revolution several other dental practices emerged. A Mesopotamian clay tablet from about 4,000 years ago is engraved with a remedy. Mix the seeds of henbane with gum mastic and apply the resulting cement to the cavity; it is not dissimilar to using clove oil. Ancient Egyptians must have been rubbing their cheeks complaining of toothache as early as 2600BCE; inside the Hesy-Re tomb complex is the earliest known individual whose tomb records the fact that he treated people's teeth.^ Around the same time, Egyptian mummies from the fourth dynasty have been found containing a wire bridge made of gold attached to a healthy neighbouring tooth. Other mummies were found to have had artificial teeth inserted in the same way as dentures. Whether these were worn daily, or even while the person was alive, is unknown.

A little more recently, about 3,000 years ago, the people of India had the rather attractive solution of serving wine at a tooth yanking, which seems a very civilized way to conduct the brutal business of primitive dentistry. And as the centuries raced forward, the technology did not improve much.

The world of dental terror is something that early humans rarely had to contend with. Impactions, malocclusions, excessive underbite or overbite, tooth decay, the violence of extractions – none of these really bothered humans until they began to plough their fields. When humans began to grow food another habit sneaked into their daily lives and in the modern world it has begun to take over.

CHANGING FOOD HABITS

The pH in our mouths changes when we eat any kind of food (with only a couple of exceptions) and some levels are much worse than others. While this may not seem only a problem for the modern human – all hominins ate, after all – the way humans began to grow food instead of hunting for it created certain conditions that we are still affected by every day.

Food was not always easy to come by for earlier hominins. One of the additional major shifts between the Pleistocene and the Holocene was the rise of technologies that began to make food both more abundant and freely accessible. This had implications for human stature and height (and much later, waists) but what it did to the ecology of our mouths is important, too. Not only were the new food sources softer (encouraging less jaw growth through lack of use); and not only did they contain higher levels of carbohydrate (which mouth bacteria feasted upon, producing plaque), but the availability of food also meant that the frequency with which humans ate changed, too. This affected them in two further ways. It extended the timespan of acidic exposure in their mouths; and it had an impact on their metabolism.

The saliva in our mouths is neither acid nor alkali, but neutral, with a pH of 7.4. One of the many jobs of saliva is to rinse our teeth and restore the correct pH to our mouths as this fluctuates when we eat. While there are obvious examples of Anthropocene foods that contribute to tooth decay and cavities – such as fizzy drinks, which can have a pH as low as 3.4 – any kind of meal will raise levels of acidity in the mouth for about 45 minutes after eating, with sugary foods exacerbating this. This phenomenon is called the Stephan Curve: the pH of the mouth dips to about 4.3 before climbing slowly back to normal. Critical demineralization of tooth enamel happens whenever the pH falls below 5.5.^

It might seem a practical response to brush food debris and bacteria away after eating, but this is a risky strategy as the brushing itself may abrade the enamel. Brushing drives the acidity deeper into the surface of the tooth across a wider area. Rather than sitting on the surface of the tooth, the acidity is ground into it by the manoeuvring bristles.

We don't know how regularly early horticulturalists ate, but we can be fairly sure that it was with a greater frequency than their hominin forbears. Today, on a normal day divided by three meals, tooth enamel is under direct attack for a minimum of three 45-minute periods if nothing is done to combat it. Add to that regular grazing and intermittent snacks, as well as the sipping of acidic fruit tea and coffee, and consumption becomes not so much frequent as something resembling a conveyor belt.

Leptin is a hormone produced by the body's fat cells and, among other things, is carried in the blood past the blood-brain barrier to the hypothalamus to signal that we are satisfied and can stop eating. When we eat, our body knows that the fat in our body will increase, so leptin is released, signalling to the brain that we can eat less and the body can freely burn its stored calories.

The obverse of this is that when we don't eat, the body dips into fat stores for energy, leptin release decreases and stimulates feelings of hunger and the body also attempts to minimize calorie use.

In the modern body, something has gone wrong with this simple system and whoever works out how to solve it is going to be a billionaire. Leptin resistance, when the brain is unable to detect how much energy is stored in the body as fat, and consequently drives appetite while minimizing calorie burn, is thought to be one of the causes of the obesity epidemic. (My guess is that the culprit is a complex polyphony of causes loosely connected to stress from certain stimulants in the urban environment, though I'm not aware of anyone currently taking bets.)

One thing we do know is that leptin adjusts to changes in habits such as meal and snack times, but we are only at the beginning of understanding leptin's role in appetite, and even then, only in the most extreme cases of pathological leptin-resistance.

Shortening the length of time between meals, while being good for advertisers and the confectionary companies, is having a detrimental impact on our teeth as well as our bodies. The feast and famine mode of consumption seems alien to us today, as this way of eating has been chipped away at since the beginning of horticulture. Our bodies have known periods of fasting for a couple of million years; we might not be used to it, but the science shows that our bodies are tailor-made to suit it.

A 2013 study, "Fasting in mood disorders: neurobiology and effectiveness" demonstrated that many "neurobiological mechanisms have been proposed to explain fasting effects on mood, such as changes in neurotransmitters, quality of sleep, and synthesis of neurotrophic factors. Many clinical observations relate an early (between day 2 and day 7) effect of fasting on depressive symptoms with an improvement in mood, alertness and a sense of tranquillity reported by patients."^

The work of Valter Longo, Professor of Gerontology and the Biological Sciences at the University of Southern California, has repeatedly demonstrated that the effects of fasting include reduced cell proliferation (turnover), reduced inflammation in the brain and gut, decreased insulin, reduced leptin, enhanced network activity in the brain, new brain-cell production, reduced resting heart rate, reduced blood pressure and enhanced stress resistance. Longo published a review with Mark Mattson, Professor of Neuroscience at Johns Hopkins University, "Fasting: molecular mechanisms and clinical applications" which concluded that "chronic fasting extends longevity, in part, by reprogramming metabolic and stress resistance pathways. In rodents, intermittent or periodic fasting protects against type 2 diabetes, cancers, heart disease, and neurodegeneration, while in humans it helps reduce obesity, hypertension, asthma, and rheumatoid arthritis."^

Mattson has gone on to research meal times. In "Meal frequency and timing in health and disease", he and his team argued that our eating habits are "abnormal from an evolutionary perspective. Emerging findings from studies of animal models and human subjects suggest that intermittent energy restriction periods of as little as 16 h[ours] can improve health indicators and counteract disease processes.”^[1]

Alongside these benefits of fasting is the fact that it vastly reduces the opportunities for pH imbalance in the mouth. And I cannot help but wonder what role feasting and fasting played in the dental health of our ancestral hominins. If you eat three times a day and snack twice and graze, then the pH of the mouth is at demineralizing levels of acidity for hours on end every day. We can only speculate about the diet of the Palaeolithic human, but it is unlikely to have resembled ours, which derive from the fact that food is only ever as far away as the nearest shop or café. There is no doubt that this changed once crops were grown and grains stored and easily accessible.

The changes caused when humans first planted seeds and harvested crops tens of thousands of years ago are still showing on our faces. Our mouths and jaws are smaller than they should be, and our teeth are so sick both of the food we are eating and the absurd frequency with which we eat it that they are flinging themselves from our skulls as if fleeing a burning building, leaping to oblivion to make room for those that are left. Through our changed eating habits, we are making it harder for our teeth to stay strong, and we are also opening the door to some pathologies by driving our metabolism into a state of constant cell renewal through regular feasting without ever experiencing famine. Perhaps as early as 20,000 years ago large groups of humans were already existing in the post-scarcity economy that many of us enjoy today.

As a result, what began with an anonymous Neanderthal and a toothpick thousands of years ago, has, in the Anthropocene, morphed into a healthcare industry worth $129 billion in the US alone, because modern humans cannot cope with the abundance in their new environment.

But settling has caused additional changes to our appearance. During this period, it is seems that every innovation made by humans has left a trail of seeds behind them. These subtle changes cause no pathology when they first appear, bringing with them only the benefits of an easier and more profitable life. But the luxury of one generation becomes the staple of the next. A simple shelter built to take refuge from the rains becomes a way of life in the Anthropocene, where for example, children sleep indoors, are driven in cars to schools where they are educated indoors, then are driven home where they

---

1 It is early days, but recent research suggests that fasting is also beneficial in the prevention of that other debilitating gateway-pathology of older age: osteoporosis. In "The effect of Islamic fasting in Ramadan on osteoporosis" published in the *Journal of Fasting and Health* in 2017 the authors conclude that it is "very effective in reducing the effects" of the disease.^

spend their evening indoors, before the whole cycle starts again the next day.

The luxury of shelter has become something resembling imprisonment. Those who built shelters during the Agricultural Revolution were planting a seed for the future that would lead to chronic pathologies, such as myopia, now suffered by hundreds of millions of people. Chronic myopia did not materialize in a population perhaps until the 19th century, and in the 21st it is expressed in numbers best described as an epidemic – but that comes a little later in our story of the Anthropocene body.

As we leave behind these most primitive early settlements, we continue our exploration of how simple and seemingly innocent choices made 10,000 years ago or more are still leaving their mark on us. The opportunity to work, to innovate, to learn, to protect is one so strong that it is blinding us and our children – all in the name of creating a better world, an easier life, with more ample resources. As we look toward the city walls being built on the sands of the Mesopotamian plains, that inability to cope with abundance only becomes more acute.

CHAPTER 4

SOIL, TOIL & GROWTH

By working faithfully eight hours a day you may eventually get to be boss and work twelve hours a day.

Robert Frost

The reason archaeologists and anthropologists can study Ohalo II is because the site was suddenly flooded and preserved 23,000 years ago. There are few human remains at the site, suggesting that as the rains fell, the group moved on to settle elsewhere, in a place which was presumably not so conducive to preservation. But as time marched on, their experiments with cultivation were part of a larger story of change throughout the Levantine region and what would later become the Mesopotamian region. Over the next few thousand years these regions witnessed a gradual intensification of horticulture and cultivation until they reached a level of intensity that could maintain entire metropolises.

The labour of these people, without help from machinery, was healthier than most of ours today. They would still have walked long distances. Later, when they reared meat rather than hunting it, we know from practically any farming community that long distances were still covered looking for grazing and pasture. Despite the fundamental change in lifestyle, we know from the bone scans of agriculturalists that life and work was filled with activity because although some work could be offloaded to animals, technology was still predominantly manual.

The Agricultural Revolution was good for much of the body (including feet and glutes) because people still had to use them. The intrinsic muscles of the foot would have been strong and tight, pulling the mid-foot into a high arch. And, just as the foot freed the hands for invention and play in early humans, out of agriculture emerged politics, art, literature – even the notion of a society. Early fashion came with the invention of fabric dyes about 6,000 years ago. The desire for colour was so intense that it emerged independently in several cultures.

In these smaller settlements other important technologies emerged such as food preservation (about 10–12,000 years ago). This meant that food could be stockpiled so that less time was spent hunter-gathering and more on other, newer occupations: tool making, the building of more permanent shelters and trade. Money also made its first appearance during this period. Before this, progress in, and the possibilities for, trade presented a particular challenge: what if your trading partner had nothing you wanted? For this reason, bartering was an unwieldy means of exchange – without a match, there could be no trade, but a primitive form of currency, cattle or certain stones, meant that trade could always take place. (Trade also necessitated the invention of writing for primitive kinds of bookkeeping, but people were not at that stage yet.)

All these changes sprang through tiny gaps in the soil. They were the hungry advancing weeds that grew alongside the founder crops of barley, wheat, lentils, peas and chickpeas – all emerged around 10,000BCE in the region of Mesopotamia and the Levant. Rice was domesticated in China, followed by other mainstays of the southeast Asian diet: soy, mung and azuki beans. In the same period, there is evidence that livestock such as pigs and sheep were domesticated, with cattle following a couple of thousand years later in Turkey and Pakistan. This did not happen only in Eurasia; we could as easily talk about the potatoes grown by the Incas in South America around 8000BCE, or their domestication of alpacas and llamas; or bananas, taro, sago and yam in New Guinea around the same period. The idea of food growth and cultivation spread around the globe like floating seeds caught on an early autumn breeze.

What did all this mean for the modern body?

With the agrarian switch, we know already that human faces had begun to change, and their teeth did not take kindly to the new chemistry of a carbohydrate diet, nor their jaws to the less challenging softer textures of ground, cooked and processed food. But there were also other, more subtle changes that took a long time to bed in to the human body. Their frames took remarkable offence at the new cereal diet, slashing their height as a species in ways so severe that some populations many thousands of years later might still be recovering.

But first we should look at the complications caused by humans living in much closer proximity to animals, what it meant for those early agriculturalists and what it continues to mean for us today.

HUMAN & ANIMAL PROXIMITY

> Whether currently-circulating avian, swine and other zoonotic [transferred from animal to human] influenza viruses will result in a future pandemic is unknown. However, the diversity of zoonotic influenza viruses that have caused human infections is alarming and necessitates strengthened surveillance in both animal and human populations.
>
> World Health Organization on influenza

In the animal kingdom, house cats are surely aficionados of luxury and comfort; they revel in it. They are sophisticates in finding that rectangle of light on a windowsill which will give them an extra few minutes of warmth, in the knowledge that when the sunlight fades, the radiator below them will click on. They engage all their senses in the predatory pursuit of comfort. On a sliding scale, humans are not far below them (think of those mattresses at Ohalo II), but the reason cats have not yet taken over the world is simply because they can't be bothered to. Getting organized is hard work and there are windowsills to occupy and keyboards to lie down on. Humans, on the other hand, throughout the last several millennia, have concentrated intently on making life as easy as possible for themselves, which is not quite the same thing.

Most technological improvement in this period of intensifying agriculture derived from the simple idea of providing food in the easiest way possible. It was a culture of "total growing" in which most resources seem to have been given over to the acquisition and cultivation of food... so that life might be a little easier. This was humans' early version of finding just the right spot on the windowsill. The history of evolution is peppered with species trying to find the easiest ways of doing just this.

As tribespeople convened and settlements expanded, growing together to meet one another like gathering clouds over a rolling expanse of ploughed fields, food was grown more and more easily, but it would be a leap to say that life was easier. New ways of life and being were certainly introduced. Through planting seeds and raising animals, over thousands of years settlements became cities and with them came more and more aspects of modern life still recognizable to us.

We need to look less at what agriculture has done for us, and more at what we did with agriculture after we discovered how it made life just a little easier, a little less precarious and a little more comfortable.

What happened when we got a little closer to nature? During the Agricultural Revolution we did not live more natural lives, but we came into much closer proximity with other species as we effectively began living alongside them. Instead of killing and eating them, we sustained them so that they could provide food and, later on, work for us. This marked a turning point in natural history, and out of it we can see nature and evolution responding not just in the reshaping of our bodies, but by changing the microbes that live in, near and around us.

The disease profile of early hunter-gatherers is lost in time, leaving no trace in the meagre fossil record we have. Diseases have to be inferred from the environment or from hunter-gathering groups today. One such disease that has a cloudy history is malaria. Its links with early human settlement are unclear, but even though the parasite responsible for the disease has been around for as long as 100,000 years its population increased dramatically about 8000BCE. Agricultural practices such as irrigation and conditions such as poor sanitation are ideal for malaria-transmitting mosquitoes. Today, nearly half the world's population is at risk of malaria, with the World Health Organization estimating that in 2015 alone there were 212 million cases and 429,000 deaths from the disease.^

Contemporary agricultural and land-use practices such as deforestation, irrigation and highland farming all disrupt the breeding patterns of mosquitoes and contribute to increased malarial transmission. Keeping any animals means that the mosquitoes have more hosts to feed on, and are also likely to contribute to the less sanitary conditions in which mosquitoes thrive.[1]

---

1 A recent study by the University of Central Florida^ (published in *Frontiers in Cellular and Infection Microbiology*) suggests that a bacteria (*Mycobacterium avium paratuberculosis*) commonly found in dairy produce such as beef, milk, butter and yogurt, as well as in vegetables grown using cow manure as a fertilizer, has a particular relationship with a couple of our genes. The bacteria is thought to activate two mutated genes in humans (PTPN2/22) which can trigger an overactive response in the host's immune system. These genes have been linked with other autoimmune disorders such as type 1 diabetes and Crohn's disease. In both cases, these are diseases in which the immune system mistakenly attacks the body's own tissue. There is also a widespread "mystery" disease associated with these genes: rheumatoid arthritis. This long-term condition can cause swelling, stiffness and pain in the joints, especially the hands, wrists and feet; it affects more than a million people in

With the domestication of animals arrived an increasing reliance on milk, which is one of the few examples of quite recent evolution in the human body. A 2008 study, "Earliest date for milk use in the Near East and southeastern Europe linked to cattle herding"^ examined the residues left on more than 2,000 pots and was able to demonstrate that "milk was in use by the seventh millennium". Early agriculturalists also found ways of processing the milk of the animals they were rearing. This food processing, by converting the milk to cheese or yogurt, reduced the lactose content and so made it easier for people to digest – this is because adults couldn't just drink the milk. Earlier humans lost the ability to digest lactose as young children; once weaned, it was no longer evolutionarily necessary.

About 5,000 years ago, with this Neolithic transition to agriculture in full sway, the people of northern Europe, from East and West Africa, the Middle East and parts of Asia, started to show an ability to digest lactose, and it became favourable for survival, at least for some populations. Once humans domesticated cows, goats and sheep, it became advantageous to be able to use their milk, especially in times of shortage and famine.

Lactase persistence is the ability of the body to continue to break down and digest lactose into adulthood. This genetic mutation was so favourable to survival that less than 200 generations later it became widespread and is considered normal, with 95 percent of northern Europeans carrying the gene. Other populations throughout the world still lose lactose tolerance in adulthood and lactase persistence is as low as 5 percent in some east Asian and African communities.

Inevitably, coeliac disease was not likely to occur until this period, either. It had no means of expression before the time at which gluten started to become a staple of the human diet. A study published in the *American Journal of Human Genetics* in 2006^ suggested that populations from about 7,000 years ago show much greater activation of the genes that help the human body cope with adverse drug reactions and carcinogens. These genes were selectively appearing in populations across southwest Asia, into Asia more

---

the US and about half a million in the UK. Like so many illnesses, it is an inflammatory condition. The study concluded that rheumatoid arthritis (although they still do not know what it is) is the outcome when an individual is born with the genetic mutation and meets an environmental trigger they might never have come into contact with before agriculture. The numbers were not absolutely conclusive (78.6 percent of people in the trial with rheumatoid arthritis showed the mutation in PTPN2), but the association is undeniably strong.

generally and across Europe, becoming active at the point at which our diet was quickly changing.

From these meagre beginnings came agriculture, and the consequences of our proximity to animals, as well as our interference in their breeding and reproductive patterns and our attempts to optimize their productivity and their bodies (remember the Anthropocene chicken?). Today, the modern human lives in a changed environment. As the population steadily grew, and people began to gather behind the walls of cities, they brought their animals with them. And as they grew in number, so did their needs, and the farms and their domesticated animals.

There was once a settlement just off the Atlit peninsula on the northern coast of Israel, which is now covered by the Mediterranean Sea. This was one of the first known settlements where people grew crops and domesticated animals. Two bodies buried around 7000BCE show a worrying sign of the speed with which an early disease went to work on its new hosts. The remains of a woman and an infant both show a distinctive pattern on the surface of their skulls, like the map of a complex delta. Analysis of the skeletal engravings (*serpens endocrania symmetrica*) shows distinctive signs of chronic respiratory impairment and failure: tuberculosis.

The intensification of the relationship between humans and animals is among the most likely explanations for the rise, spread and mutation of diseases such as tuberculosis. Today TB kills a massive 170 million people every year, with the AIDS crisis in Africa exacerbating its virulence. Known in the 18th and 19th centuries as consumption, this disease does indeed eventually consume the sufferer. It is contracted via inhalation usually after sustained exposure to the *Mycobacterium tuberculosis*. This is a slow-growing bacterium which mainly attacks the lungs, but it can affect any part of the body, the stomach, bones or the nervous system. Eighty percent of TB cases today are in low- and middle-income nations and it is believed that the disease was almost non-existent in humans before agriculture. As to whether it originated in cattle and was passed on to humans; this is unknown. If it was, it was likely that new forms of human and bovine proximity proved to be the cause. Once agriculture was industrialized, the mathematics of chance started to play a role in the likelihood of contracting the disease.

There are tens – or maybe hundreds – of diseases not unlike TB that have almost certainly emerged from agricultural practices, whether from animals themselves, or through their being the conduit that allowed humans to gather in ever larger groups.

Now there are 1.5 billion cows on the planet, each one producing the bacteria *Mycobacterium avium paratuberculosis*, which is widely distributed through meat, dairy and via soil-enrichment to vegetables. This makes the bacteria impossible to avoid, creating an ideal environment in which the mutated genes can take up their cue as eagerly as a bit-part player on the opening night of an am-dram extravaganza.

Tuberculosis is a "crowd disease", one that needs a large number of people to flourish. The population size of foraging groups was too small for infections such as measles, mumps and smallpox to spread. Today, the number of animals (pigs, cows, chickens and turkeys) that humans produce annually is so high as to be practically meaningless. Tens of billions of chickens alone are consumed every year. We may as well try to count the number of stars in the galaxy, or the grains of sand on a continent. On one hand, there is a minuscule likelihood that a new mutation of a virus will emerge in a pig or chicken. But that same minuscule number multiplied by billions of throws of the dice (the number of animals reared, slaughtered and processed annually) changes the likelihood of a new virus emerging from a tiny "if" to a "when". Animals are now reared with such intensity that mathematically it is only a matter of time before one of the many mutated flu viruses becomes an epidemic that passes freely to, and between, humans.

Farming animals no doubt provided us with opportunities for survival and growth, but with the intensification of farming practices today which encourage food-borne illness and antimicrobial resistance, the scene is set for viruses to mutate, trading genes to become the next super-flu transmissible between humans. There are major flu outbreaks approximately every three decades. We are currently overdue a visit from one.[2]

This is our inheritance. This is what we have done with agriculture; but it is not yet done with us. During that period there were other changes and we are still experiencing these in our bodies today.

2 This is just a mathematical assumption based on the fact that there have been nine major flu pandemics in the last three centuries. The last one to kill over a million people was half a century ago: the 1968-69 Hong Kong flu. Before that, there was the 1957-58 Asian flu which killed about 2 million, and of course, the Spanish Flu of 1918. In recent years, the "swine flu" of 2009, claimed less than 20,000 lives in total.

Cosying up closely with animals and crops had another dramatic effect on our bodies, and even though this story spans the entirety of the genus *Homo*, the action mostly took place during the Agricultural Revolution. I am referring to height. During the last 200 years we have suddenly started growing as a species at a rate not seen before in human history. Are we taller than we have ever been? Are we likely to keep on growing, and how tall will we get? The answers to these questions are all tied to our relationship with those crops and cereals that filled our stomachs… but not our bones.

THE INCREDIBLE SHRINKING MEN & WOMEN

With the alacrity of a gifted croupier, the Anthropocene human can readily recite a list of ways in which we are seemingly much healthier than our ancestors were. At, or near, the top of this list would be: "we live much longer" and "we are much taller". In this section we will take a first look at the fact that there's a lot less to these claims than meets the eye.

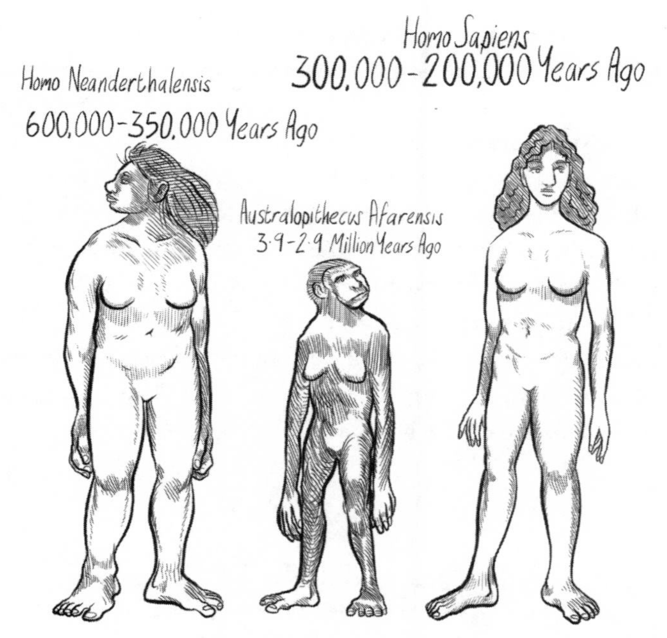

There is considerable variation to be found in the heights of some hominin species, with *Homo erectus* being consistently the tallest.

We will leave longevity on the bench a little longer and look at the astonishing impact agriculture has had, not just on our faces, our teeth and our susceptibility to the diseases we had a hand in creating, but on our height.

The modern Anthropocene human is noticeably taller than humans used to be. While there are substantial variations across time, place and even different kinds of human, there is no doubt that we are growing longer bones that are not seen in the fossil record. Unlike our growing feet, no biomechanical trick is fooling us into believing that we are of greater stature. We bang our heads on the beams of ceilings in working men's cottages from the 16th century because they were built for people much shorter than we are today. This begs the question: how big are we going to get? Will we all be ducking under basketball hoops in a century or two, or is the story of our growth nearly over?

The relationship between our height, agriculture, the move into more intensive farming practices and the beginnings of metropolises is more than a little complex. But the earliest human females for which we have fossil evidence averaged approximately 125cm (4ft 1in), with males about 157cm (5ft 2in) in height (*Homo habilis*). In Lucy's species (*Australopithecus afarensis*), females averaged around 104cm (3ft 5in) and the males 150cm (4ft 11in). *Homo floresiensis* were about 107cm (3ft 6in) tall. The species has only been found on the islands of Indonesia, and a common hypothesis about their size is that it is relative to population and geo-resource density, which are inevitably more limited on smaller islands. There is something strange about the *Homo* genus, because *Homo floresiensis* was about 99cm (3ft 3in) tall, while the man who used to work in the next office to mine was well over 198cm (6ft 6in). That is not simply variation.

The story of *Homo sapiens'* height is more complicated than that of preceding species, mainly because there has been so much DNA and so many environments for it to express itself in. About 10,000 years ago, males were typically about 163cm (5ft 4in), and many believe that their rather diminutive size was a response to the colder climate of the Pleistocene, the difficulty of growing crops successfully and consequent malnutrition. People certainly ate a more restricted diet, but the chart of our species' average height through the ages does not rise neatly or slowly. It is more like the upward arc on the world population chart that rises explosively in modern times.

Between 10,000BCE and the 17th century, when all manner of technologies were refined and food provision techniques had had many millennia to mature, what was the net increase in human height?

Two and a half centimetres (one inch).

Nearly 12,000 years of abundant food yielded a single inch of growth. In the 18th century, the average height of an English male was 165cm (5ft 5in). Now it has leapt to 178cm (5ft 10in) and it seems still to be climbing.

The anatomically modern male humans who first arrived in Europe 40,000 years ago had a mean height of 168cm (5ft 6in) – taller than 18th-century men. There is a simple correspondence, though, between height and environment. Being shorter is much more efficient in colder climes. A longer length of bone, on the other hand, makes you much more efficient at losing heat (especially during endurance hunts on the savannah). So, understandably, the European male in his cooler clime shrank a little.

By the Late Upper Palaeolithic era (8000–6000BCE) average height had dropped by about 13cm (5in). Height further decreased among some cultures to below 152cm (5ft) during the Late Neolithic (5000–3000BCE). According to Michael Hermanussen in his study "Stature in Early Europeans" the "body stature of historic Europeans remained within the range of 165–170cm [5ft 5in] for males, and 155–160cm [5ft 2in] for females up to the end of the 19th century."^

The story is about as confusing as one could imagine, with many influencing factors from tribes, societies and cultures, all in play at any given moment. Even the data from deep history is skewed because those preserved – those buried – tended to be the most revered and affluent, and they are likely also to have been among the best nourished in any given group. The common thread is that with the intensification of agriculture, especially around larger early settlements, diet did not improve, but worsened, and this had a marked impact on stature. The foods eaten changed, from meats with high protein content and rich in nutrients, to cereals with much less protein which were less nutritious.

Geneticists understand height quite well. They can make sense statistically of the range in height of the offspring of a male and female. Thanks to the tireless experiments of those early 19th-century geneticists such as Gregor Mendel and Francis Galton. The latter – a big believer in the numbers – concluded that tall parents tended to have taller children but not absolutely, only on average when compared to the rest of the population.^ The range of heights among the offspring of taller parents showed the same bell curve found among any parents' offspring; some were a similar height to a parent, some shorter, some taller. But there are many factors involved in the determination of a human's height.

Epigenetics is one of the most important factors in understanding the explosive increase in height of the Anthropocene human. Epigenetics is the study of gene expression; it looks at how genetic change is induced or affected by environmental triggers. During the last couple of decades, the number of journal articles devoted to the subject has risen from a handful in the late 1990s to thousands every year. Despite this volume of study, there is still a great deal to learn about its complexity.

There is DNA in most cells in our bodies, but only some of our genes need to be switched on in any single microenvironment. The Hox sequence, for example, which instructs the placement of limbs, heads and tails, is not necessary in most other parts of the body of an adult so needs to be switched off in organs making digestive enzymes, for example. For this reason, each cell only turns on a tiny percentage of its genes; the remainder are repressed: switched off, through a process known as gene regulation. A more lasting "locking" mechanism for gene expression is called methylation. Sometimes this activation and deactivation happens very quickly indeed, for example, when you eat food and it is digested – and sometimes this process of methylation can last a lifetime, or even longer. The changes wrought by methylation are sometimes simple but can also be complex and vast.

In terms of our height, roughly around 80 percent of it is determined by our genes. A study conducted in 2010 used the genetic data of 183,727 individuals to "show that hundreds of genetic variants, in at least 180 loci, influence adult height."^ At best, there is only partial understanding of how these genes work together to determine the final height for the organism. There are some genetic mutations that have a huge impact on height.[3] Most other genes, though, have only a modest impact on final height.

There is an undeniably strong argument for genes and height. But while that genetic determinant might seem to show that our height is pretty much set, it actually means that our environment might determine whether we are going to be, say, 152cm (5ft) or 183cm (6ft) tall: quite a big difference! Our genes intend us to be a certain stature, but if they don't meet the right

3 Such as the FGFR3 gene which causes dwarfism, when the body typically has a normal-length torso with shortened limbs.

environment at all the key moments of development, they can't make the body grow as tall as they might like, and our height, once set in adolescence, stays the same until old age gradually shrinks us.

Diet and height are connected. When Alice in Lewis Carroll's 1865 *Alice's Adventures in Wonderland* first tumbles down the rabbit hole, she eats a currant cake and suddenly telescopes in height to "more than nine feet". Carroll was onto something: food translates into height, sustenance becomes stature, grub begets growth.

There have been countless studies of diet and the ways in which it has changed in certain populations over the millennia. One distinct group of people is, at least in European terms, a relatively remote people; not isolated, but through habit, geography and also temperament and disposition. They are unique among other populations for two reasons: they live a really, really long time, and they are very short. They have also had a difficult relationship with the agricultural diet.

THE SHORT STORY OF SARDINIA

The history of the Sardinian body is one which illustrates more broadly what happened to the bodies of Europeans when the change in agriculture began to take place. The reason Sardinian morphology is worth looking at is because we can see the story of human height close up, focusing on the similarities and differences of a specific population, and especially their differences from other Europeans. But it also tells us that the spread of agriculture was slow and not merely affected by new habits and technologies which spread with the pace and unrelenting persistence of water poured onto hard ground.

Agriculture needs the right environment to succeed. In some places this was easy – especially in verdant flatlands. In less accessible places, without wide open fields that could be planted and ploughed, agriculture took a lot longer to put down roots – thousands of years longer. The changing height of the Sardinian body tells the story of how agriculture changed the Sardinians, and consequently, all of us. While the story of our height more generally spans many millennia and takes place across countries and continents, the Sardinian story is one that subscribes to Aristotle's dramatic rules for a really good drama: being a story that unites place, time and action.^ The story of height globally is one of epic scale; in Sardinia it becomes a concise, short story.

Dr Gillebert D'Hercourt was probably the first serious anthropologist to wash up on the shores of Sardinia. In 1850, he published a report in which he described the Sardinians unflatteringly as a group of pygmies.^ Anyone who has travelled to Sardinia will have noticed their shortness. And although it is less evident in the island's cities (such as Cagliari and Oristano), a brief visit to one of the mountain towns or villages will attest to the fact that Sardinians are shorter than other Europeans and Americans, but that younger generations are without a doubt taller.

A recent study described analytic research into the history of Sardinian stature, what it means and what it can tell us about the agricultural diet.^ The authors compiled data that tracked the relative heights of Europeans and Sardinians during different periods starting at the time of the Agricultural Revolution.

As the first seeds of agriculture began to drift across Europe about 6,000 years ago, people who at the beginning of the period were 170cm (5ft 7in) tall dipped nearly 5cm (2in) over the following three millennia, before slowly regaining lost stature during the late Victorian period and then climbing quickly during the 20th century.

The Sardinian story in statistical terms was dramatically different. Fossil remains of Sardinians from 6,000 years ago show that they were about 9cm (3.5in) shorter, then the mean suddenly climbed to meet the European average during the Bronze Age. There is no obvious reason for this sudden climb. The authors of the 2016 study suggest that it "could be attributed to the coexistence of different ethnic groups on the island or, alternatively, to substantial disparities in the living conditions of early settlers" with populations in the more accessible northern regions of the island showing more growth than those in southern and central areas.

The real story started after this growth spurt. During the years that followed, when Europeans were working hard to reclaim that single inch, Sardinians' height dipped until the middle of the 19th century. Then between 1861 and 1874 it dipped sharply, so that Sardinians in the 1870s were, astonishingly, shorter than those early settlers in the Neolithic. Another shocking thing about the data is that after this point, Sardinian mean height began to rise at exactly the same rate as it did for Europeans.

Genes did not suddenly change in 1874. There had to be a social, political or cultural cause for the change – and it must have been potent to have had such a strong and sudden impact.

It was, in fact, that the lag in height was the result of the agricultural diet arriving late in Sardinia.

The Nuragic people of prehistory (Sardinia's first settlers), have been shown to have lived predominantly on animal-derived proteins such as meat and milk. With the arrival of agriculture, this diet shifted to grains, starches and carbohydrates. The less affluent ate a predominantly plant-based diet. The study authors explain: "Global health and body height were greatly affected by these changes that lasted well after the medieval period into the 18th and 19th centuries."

They are able to tell that diet changed the stature of Sardinians because of leg length. The growth plates in the legs are most active during adolescence and poor diet alters the rate at which extra bone can be made, so that what you eat at certain points during your development determines to some extent how tall you can become. The intensification of agriculture in Sardinia in the late 19th century had a global impact on its entire population, as did diseases such as malaria, helminths (worms) and other infections. And as Sardinians were growing taller, their legs (rather than their torsos) lengthened, suggesting strong links with diet.

In comparison, at more or less the same point in history in Britain, the story was a little different. Diet was changing because a much greater variety of food became available. Conditions were always worse for the poorest, but even those on only modest incomes in the smaller towns and villages found themselves more connected to the world around them, with an accompanying growth of variety in goods and supplies.

These revelations appear to make the source of our stature clearer. If we are well fed at the right times, our genetic potential for height should be met. But the question stands, why can someone as seemingly short, at 172cm (5ft 8in), as I am, see over many of the heads of Sardinians? Being short shouldn't change your DNA; it's not the same as the effects of smoking or extreme age.

I don't think this to be the case because my legs are not short for my height, but if my DNA had really determined that I should be 183cm (6ft) tall as my brother is, for whatever reason I didn't have the right amount of proteins, calcium and so on at the right time to grow to that height, then surely I should pass on my height genes to my offspring? My genes should produce children in a bell curve around 183cm (6ft). Moreover, because I had done all this research into diets and stature, they should, I would hope, grow to their full height. The whole matter of slightly-stunted growth should be cleared up within a generation. But height doesn't work like this. We know from data that short populations, on the whole, remain short.

Sardinians may track the same curves of increasing growth that the Europeans are enjoying, but they are not catching up with them – they are just following a similar trend from a lower baseline.

One of the ways we might understand this is through a process called "epigenetic assimilation". Barry Bogin, Professor of Biological Anthropology at Loughborough University, explained "a new hypothesis, which focuses on those epigenetic processes regulating gene expression, metabolic function, physiology and behavior."^ This means that methylation in our DNA (the on/off switches) may be passed onto offspring, activating a similar restrictive growth strategy in the DNA of the offspring as being more favourable to survival in the supposed harsh climate.

Methylation is influenced by many factors and nutrition is a significant one. What is surprising is that in some cases (and not only our height) these changes can be handed down through the DNA for generations to come. Bogin has also argued that along with environmental conditions that might affect food quality and variety, political upheaval can have an impact on stature too. He has cited a drop in height of black South African men between 1900 and 1970, which was probably caused by apartheid. This can become a downward spiral. Bogin told *Live Science,* "It shows you the power and how the generation after generation effects of something bad that happened to your mother gets carried onto you and your children, and it takes about five generations to overcome just one generation of starvation, or epidemic illness, or something like that."

One of the other factors is, of course, sexual selection. For years I'd assumed tall people had all the fun and so were more likely to produce taller offspring and that would result in a taller population. I now know this is not the case, but there have been some interesting studies that explain how evolution holds the reins on this idea of exponential growth in stature. The overall height of our species might be grounded through some subtle evolutionary mechanisms.

A 2012 study collated data from 10,317 women and men^. The team focused on sets of female and male siblings, looking at their height and their relative reproductive success. The study cautiously concluded that the taller men had more children, but the heritable genes for height were held in check by the fact that shorter women were comparatively more fertile.

Does the common assumption that tall is better bear scrutiny? Is reaching our full genetic height really so desirable?

On the one hand, being taller has been consistently statistically linked with earning more. This will probably lead to a better, more affluent lifestyle with greater potential for accessing important resources and good palliative care in old age. Thus, being tall does have its advantages.

On the other hand – and this may sound like sour grapes from someone who is 172cm (5ft 8in) – the Sardinians certainly have not reached their full genetic height and some of them are among the longest-living people on the planet. The research is in its early stages, but there are strong associations between height and longevity, and not just in Sardinian populations. Some regions of Japan (especially Okinawa), boast a high number of centenarians, and in a nation not known for producing people of taller stature, Okinawans (where men average 145cm/4ft 9in) are shorter than their mainland counterparts by several inches.

This population research is starting to look at another age-old assumption: that women live longer than men. Instead of this being a mystery governed by gender, some researchers now think that women's longevity may be a by-product of their generally smaller size; that their lower metabolism is the principal contributor to their longer lives.

The Sardinians who live in mountain villages are hard grafters, daily working their flock and farming well into their seventies. Agriculture may have taken a long time to arrive there, but the geography of the landscape meant that it was never going to dominate it easily. So travelling to these places is like stepping back in time, before much technology, and before many of the comforts we associate with city life existed. While the mountain villages in Sardinia may have grown in the last few thousand years, they have not done so exponentially in the way that the first cities did. Sardinia was (and to some extent still is) a world away from Ancient Rome or Greece, Babylon or the ancient city of Uruk in Mesopotamia.

CHAPTER 5

EXERCISE, ERGONOMICS & LIFE & DEATH IN THE CITY

He looked at the walls and was awed...

The Epic of Gilgamesh

Mesh-ki-ang-gasher of E-ana, son of Utu, reigned for 324 years
Enmerkar, son of Mesh-ki-ang-gasher, who built Uruk, reigned for 420 years
Lugalbanda, the shepherd, reigned for 1,200 years
Dumuzid, the fisherman, reigned for 100 years
Gilgamesh, whose father was a spirit, reigned for 126 years
Ur-Nungal, son of Gilgamesh, reigned for 30 years

Excerpt from the Sumerian King List

Some time during the 1850s, archaeological work was conducted just outside Mosul in what is now Iraq. Locals dug and loaded donkeys' panniers with hundreds of bits of broken clay tablets found at the site, an ancient version of a library at the Babylonian city of Nineveh. Most of the tablets were badly damaged, but some had enough information on them for their meaning to be decipherable (if you were adept enough to recognize an alphabet of up to 600 characters that could have been written in any number of about 13 long-dead languages, with no vowels, no punctuation and no word or paragraph spacing). The library was about 2,500 years old, but because the letters had been imprinted on moist clay with a sharpened reed and then baked, the texts had been preserved in a way that they could never have been on paper. And just as a modern library might contain works by Shakespeare or Homer, Ashurbanipal's library at Nineveh also contained many older texts.

The bits of broken clay undertook a slow and arduous journey all the way to the British Museum in London where they were unloaded. Only a handful of people were able to read the cuneiform inscribed on them, indeed a couple of decades or so later they were still being deciphered.

George Smith must have been the single most highly-skilled "temp" who has ever lived. His interest in this ancient script was so acute that it spurred him on to learn more about it until he reached the point when he was one of the few people in the world able to read it.

One day in 1872, Smith was processing and cataloguing his way through the pile of broken bits at the museum in London when he hit the jackpot. He came across a shard of clay, perhaps once a little larger than a postcard. On it, he read of a great flood, of a man who had survived it and sent out a dove from his boat in the hope that it might find land, and of the dove's return with an olive branch.

Smith had discovered an ancient Mesopotamian version of the story we know today as Noah's Flood, probably dating back to a couple of thousand years before the Bible, from *The Epic of Gilgamesh*.

The story was gradually pieced together, starting with Smith's trips to the sites in the 1870s (before he died suddenly in 1876 while trying to make it home to his family). And since then, more and more fragments of the story have turned up.

The story is set around the third or fourth millennia BCE and recounts the adventures of a real king from the time. Gilgamesh starts the story as a vain and selfish ruler. The gods decide to punish his hubris, so they make a fearsome wild man out of clay to fight him. Instead, the two become friends and go off on a boys' bonding trip to the forest to collect cedar wood. In the forest, they meet Humbaba, the trusted protector of the forest, and they kill him. They also destroy the entire woodland area. The gods are outraged by the outcome of their plan, so decide to punish Gilgamesh by killing his new friend. The king is heartbroken at the loss, and hearing of an ancient man who had survived the flood, goes to find him to discover the secret of his immortality. He finds this Mesopotamian Noah who tells him where to find the plant that he must eat so he can live a long life. Gilgamesh goes off in search of it and finds it, but later while he is bathing a snake slithers up and steals it. Gilgamesh returns to the new city of Uruk a changed man. Sad at the loss of his friend, and with the knowledge that he too will one day die, he sets about finding immortality by building some great walls to protect his people, walls so great that generations to come will see them as a symbol of his immortality. This was probably the first time in history, but not the last, that a ruler attempted to galvanize people with the promise that he would build a wall.

The Epic of Gilgamesh is a colourful, accomplished and moving story, and it tells us a great deal, not factually in a way that the archaeology might, but about the structures of this early city-society, its values and how it might have worked.

The Mesopotamians had writing for well over a thousand years before the Greeks, and this epic tells us about the ancient city of its setting, as well as speaking eloquently about the politics, the customs and assumptions of the very earliest city dwellers.

These were the people who built and developed settlements, transforming them from simple village arrangements like those at Ohalo II into mega-cities, fortified with walls and with populations of tens of thousands, rather than just a few families.

There were a handful of early cities scattered about southern Asia and the Mediterranean. There were settlements in Mesopotamia, Minoan Crete, Egypt and one from about 2600BCE in what is now Pakistan, called Mohenjo-daro. At its height, it had a population as large as 50,000 and would have come a close second to Uruk in size.

The world the poem describes is more similar to our own than might appear at first glance in the following ways. Uruk had a leader (the king), but also a council of elders who expected Gilgamesh to behave in a manner that was honourable and just in his treatment of his subjects (which meant not sleeping with other men's wives on their wedding day, for example). The religion of its audience looks quite similar to those which emerged much later in Greek culture: polytheistic, with gods who did not always have the interests of the populace at heart. The humans who died went to another world where they continued some sort of existence. The gods seem as patriarchal as the people of the city and were quite human in their ways: they mated with one another, had offspring and were interested in influencing others (especially the humans).

What we also see in the poem is that there were substantial differences in the relative wealth and power of the people of Uruk. The world depicted in the poem was a "man's world"; earthly women were practically non-existent, presumably subordinated to the domestic sphere. The men were inclined to war and praised for their inclinations; their strength was admired, as was their fighting ability.

Between the settlement and the city, inequalities inevitably became more prominent and we can only guess at how much work people did, especially at the lower end of the scale – certainly King Gilgamesh was not overwhelmed by answering emails when fulfilling his role as ruler of the known world; but the farmhands and the protectors who sat on the city walls were unlikely to have enjoyed favourable working conditions while they defended the comforts of others.

In those early cities, people settled and slept in permanent residences; they cooked together and were more likely to draw water than to travel miles to find it. Just like any city, ancient conurbations like Uruk in Iraq began by being temporary, and then became small settlements, and grew over the course of hundreds of years into urban centres. Once there were cities, there were also royals, nobles, priests; there were classes. Animal territory transmogrified in humans into property ownership. Merchants would have employed staff, scribes would have had tutors and accountants, architects and shipbuilders would all have enjoyed good lives in these early cities.

The "working" class would have consisted of potters, blacksmiths, carpenters, brick-makers, brewers and taverners, fishermen, butchers, builders, weavers, basket makers, charioteers, sex workers and military men. The gender pay gap in early Mesopotamia was less pronounced than one might expect. Taverners and brewers were often women, and they mostly enjoyed equal rights under the law (though no access to an education).

Among the substantial collection of tablets from the British Museum is a very early payslip (about 3000BCE) indicating how much beer is to be given to a worker for duties performed. This represents a substantial social shift from earlier settlements, which would have been founded upon sharing, to a more modern society that had clear hierarchies: employer and employee.[1]

Cities also gave rise to slave ownership. People who were captured in war (which was very common) or bought and sold in the market became

1 An interesting parallel emerges here. The habit of paying in beer did not exactly persist, but during the 19th century workers met in pubs to collect their weekly wages. The landlord and the boss had come to an agreement that a tab could be run, with the boss deliberately showing up very late in the evening to clear the debt and give any change to the drunken, out-of-pocket workers.

the property of their owner. Many did similar work to those above them in the social strata.

From these new city streets came the wheel, the plough, the idea of a code of law. We still use Sumerian mathematics when we count up seconds into minutes and minutes into hours (base 60). This sexagesimal system also survives in the degrees in a circle, or the inches that make up a foot. But then, I suppose it makes sense, the average height of a city dweller would probably have been 152cm (60in).

In early cities, people often ate fruits and vegetables, but everyone ate barley all the time. Some meat, fish and eggs were eaten and people regularly used flavoured condiments, salt, oils, herbs and spices. They drank wine, too. Other cuneiform tablets revealed that the inhabitants of Uruk were already enjoying the habit at the time of the real Gilgamesh's reign.

Although there is little evidence, there can be no doubt that for the masses of Uruk existence was something to be fought for, by all those who reared flocks, baked as they worked in the desert heat, tilled the soil, managed crops, fought regular wildfires and dammed rivers to protect against the deluges that frequently flooded the plains (hence the Noah story). Life expectancy must have been at limbo-low levels, and only a fool would call this in any way a "natural" order. Has there been a time or place in history when the exploitation of resources has not been equally reflected in a similar exploitation of women and men?

An obvious question arises when we look at the rise of the city-state. Except for the slaves who had no choice in the matter, why did the masses willingly surrender their freedom and agree to produce an excess of food so that others could live more freely than they? Why did they allow those early politicians and bureaucrats to enjoy lives of relative leisure? The answer surely must have been the fear of war.

As settlements spread throughout the region, the warring among tribes would have been ceaseless and costly. Women and men would have lived with a constant fear of assault or attack, knowing that at any moment they or their loved ones might be killed. Trading this fear for the extra work that would result in metropolitan life does not seem too difficult a decision to have made.

Although *The Epic of Gilgamesh* is what we would now call fiction, we know that a King Gilgamesh is mentioned in the Sumerian King List, which dates his reign as lasting for over a century around 2600BCE. Long life was a badge of honour for the rich.

The remains of Uruk are still there in the sands of Iraq. The remains have a lot to tell us about the way those Sumerians began to live, how their appearance changed – and consequently, ours too. And they also tell us how, more than anything in the world, what they craved was something they believed no riches might buy: more life.

THE HEALTH OF CITIES OLD & NEW

The first tale of literature and poetry as we know it is that of King Gilgamesh, who wanted to live forever. Immortality is of little appeal to most, but it seems to be a human trait to desire more *good* life. Because Gilgamesh is right there at the beginning of human literacy, it is not possible to say what it was about city life that gave him such a keen desire for longevity over all other things. Living in Uruk, one must imagine that most folks' greatest desire would have been a table full of food, a jar overflowing with wine, but what is interesting is that King Gilgamesh already has all these things, so his desire naturally extends to what money and power seemingly cannot buy: immortality.

If life expectancy in Ancient Rome was anything to go by, city life was very cruel indeed – at times it dropped to as low as 19. All the diseases that occur in environments with poor sanitation or where large numbers of people live in close proximity, thrived in the city. Was more life craved by Gilgamesh precisely because existence was even more precarious than it had been out on the plains? We see life better when death casts its shadow in the doorway.

How long are we supposed to live? What is a normal life expectancy? How long did people live before they began piling into the first cities?

Data for life expectancy in Uruk is not easy to come by. There the elderly were generally respected, and as in hunter-gathering groups they took pleasure in a reputation for knowledge and wisdom. Age prejudice did not exist in foraging groups as it did in urban ones. Among hunter-gatherers the elderly continued to be useful well into their late years. Yet, in a collection of *Babylonian Wisdom Literature*,^ there is among the proverbs the declaration of a sex worker bemoaning, not her age but its social perception. She tells us, "My vagina is fine, (yet) among my people it is said of me, 'It is finished with you.'" Age prejudice certainly existed. Urban life had less need of the kinds of work that can be performed in old age. What the average age of the people of Uruk might have been is harder to calculate. From skeletal analyses of specimens from around 3000BCE, just less than a third made it to 60. Specimens from Kish in the Persian Gulf suggest an average life expectancy of about 30, with only about a fifth of people living past 35.

Life expectancy for early humans was about 25 years old. This is the single most common reason for rejecting ideas that look back at previous incarnations of humanity for guidance on, for example, how we might live well for longer today.

Out on the savannah, humans were regularly exposed to substantial risks. The world was changing quickly and drought or famine could strike at any moment, but despite the life expectancy, very few early humans died in their twenties. While the risks incurred from violent encounters with predators, prey or other tribespeople were significant, 25 as a number is utterly misleading.

Today in the UK and the US infant mortality rates are well below 1 percent, so to speak of an average (or mean) age of death makes perfect sense. Those numbers really do tell us how long on average the people in that community live. But for hunter-gatherer communities we need a different mathematical model because without healthcare, infant mortality rates could be in the region of 40 percent. If nearly half a group dies in infancy and the other half all live well into their early fifties, that yields a mean of 25: misleading at the very least. Life was hard, but it was not as short as the numbers press us to think. If you were a hunter-gatherer and you made it to adolescence, there was a strong likelihood that you would live to be 60 or 70 – not so different from modern humans.

The relationship between civilization and longevity is a complex one and once the figures are teased out, the huge gap in supposed life expectancy narrows significantly. The Tsimané, for example, are an indigenous forager people of lowland Bolivia and their modal age of mortality is 78 (being the data point that appears with greatest frequency). Higher-income nations, with advanced healthcare and a much improved diet, have a better figure for modal mortality, but instead of the 50 years' difference we hear bandied about, it is closer to about seven years. The new city dwellers still live longer, but by nowhere near the margin that average life expectancy might lead us to believe. Everyone knows that neither group is experiencing optimum conditions for longevity. The Tsimané would benefit from greater variety in their diet and access to healthcare; and city dwellers in higher-income nations would benefit from lifestyles that relied less on processed foods and refined carbohydrates, encouraged more movement and had cleaner air.

Height and life expectancy are statistically linked. Shorter stature, especially in people who live in built-up communities such as cities, is often indicative of a sparsity of local resources but is also associated with several diseases that can have an impact on lifespan.

In search of the reasons our great, great, great ancestors sought longer lives at such an early stage in their evolution, I wanted to find out how the difference in modal age could be so narrow when their lifestyles were so different. There are a few places on the planet that show a happy middle ground between these two extremes of lifestyle. These areas of peculiar longevity are certainly not high-income regions, and they are home to more centenarians per capita than anywhere else in the world.

I decided to go to one to meet one of the authors of the Sardinian stature study, a man famous for his work on the staggering longevity of people in regions that boast almost no healthcare, have no nursing homes and possess very few of the conveniences of modern life.

THE LONG STORY ABOUT SARDINIA

When I told a younger colleague that I was doing research into why some people live beyond a hundred years, he exclaimed something along the lines of, "How awful; why would anyone want to live *that* long?" It is a common sentiment that Jonathan Swift's *Gulliver's Travels* from 1726 exploits nicely. In the third part of the novel, our travelling hero, Gulliver, visits the Isle of Luggnagg where he meets the Struldbrugs. These inhabitants seem like normal humans except for the fact that they cannot die. Unfortunately for them, though, all the other signs of ageing proceed as normal, such as hair loss, shrinkage and worsening eyesight. Gulliver tells us:

> As soon as they have completed the term of eighty years, they are looked on as dead in law; their heirs immediately succeed to their estates; only a small pittance is reserved for their support; and the poor ones are maintained at the public charge. After that period, they are held incapable of any employment of trust or profit; they cannot purchase lands, or take leases; neither are they allowed to be witnesses in any cause, either civil or criminal^.

They are useless members of society; worse, they are a drain upon it. Their levels of exhaustion make them unhappy. The promise of release from their torture, namely death, is even denied them.

Is this what living longer means? The gradual onset of decrepitude, uselessness, with the added twist of lots of pain, discomfort and morbid diseases?

Perhaps the world's leading specialist on ageing in Sardinia and its seemingly slower processes in the region, is Gianni (Giovanni) Pes, who works at the University of Sassari. He has been studying and experimenting in the field for more than two decades. When he and a colleague (Michel Poulain) first went into the field to identify areas of peculiar longevity, they did not have computers and GPS link-ups.

Once they had corrected for certain social factors that affect life expectancy (namely, extreme wealth in places such as Monaco where residents can afford extensive palliative care) they located several regions of longevity, one of which was Sardinia. Armed with a map and a blue felt-tip pen, they worked their way around the central parts of the island, identifying these places and circling them on their map. They became "blue zones".

The more I learned about longevity, the more I could see that it was tied up with questions about our historical relationship with agriculture, and that at one level it revealed that the quality of food that farming produced, while keeping our bodies active in the production of it, was not always ideal for a long, good life. So, what was it about the lifestyles of those in some of the less-populated regions of Sardinia that blended modes of living, both ancient and modern, in ways that have fascinated and perplexed travellers for generations? I was determined to find out more about them, so I arranged to meet Gianni Pes.

Pes's initial interest in longevity derives from the fact that his great uncle lived to be 110 years old – at the time, he was one of only four men in Sardinia to have reached such an age, but this number has since more than doubled.

The idea that longevity is determined by our genes has come under increasing scrutiny in recent years, just as height has. A famous study of Danish twins^ concluded that "longevity seems to be only moderately heritable". But whereas up to 80 percent of height might be genetically determined, only about 25 percent of longevity is governed genetically; the rest is environment.

This means that most of the factors giving us a longer, more actively-engaged life are cultural and environmental. The important thing to remember is that these extra years are usable. They are not an additional decade or two spent in the corridors of death in a nursing home. The years Pes is interested in adding to everyone's lives are active ones in which people choose to continue to work and lead the life they are accustomed to living, one in which they are comfortable.

I wondered how his research might have changed his own behaviour. Was there anything he now did every day as a result of it? "One of the most important aspects of this lifestyle is the levels of physical activity. These people in blue zones do not live near their work place. In Villagrande, for example, the average distance walked every day is 10km (6.2 miles), and it is a very hilly area, which makes the activity more intense."

Pes explained, "I wake up early, about 6am, and I walk 7km (4.3 miles)." He then joked, "Ten years ago my belly was a little round, like you more or less, but now, after doing this activity for a few years (he stood up and gestured to his undeniably flatter stomach) I feel much more ready to work in the morning." I asked him if he always manages this without fail. He turned to his wife, and after a quick exchange they established that he had only not gone for his daily walk about two years ago, when he had had a fever. Walking distances comparable to those walked by our ancestors is the first key factor.

"The second thing is to be more attentive to diet." Pes's general theory is that there is no perfect diet; he is rightly sceptical of the diet industry. Nonetheless, "I try to avoid simple carbohydrates like white bread. I eat a little cheese every day; red wine also seems like it might be part of the explanation for longevity in the blue zones, as they contain proanthocyanadins." These chemicals bond with collagen, helping to protect the skin (especially from UV damage). They can also help with flexibility in the joints, and improve blood circulation by strengthening artery, vein and capillary walls.

Like gravity, nobody really knows what ageing is. We can watch an apple fall from a tree and learn everything about the behaviour of gravity, its processes, its inevitability, but what it is, is much harder to determine. Ageing presents us with similar difficulties – everyone recognizes it, what it does, what its signs and processes are, but no one can really explain it. As Dan Buettner says in his book *Blue Zones*,^ we know that it has only an "accelerator pedal" and we have quite a bit of control over how much gas we give the engine. One of the theories of ageing is that it is essentially a form of genetic mutation, which

Pes and others believe has strong links with diet.

"Probably a sedentary life in itself is not dangerous. It becomes dangerous because it increases the chances of developing metabolic diseases. Also, an excess of adipose tissue (fat) is a risk, too. The fatty tissue that we have under our skin on our arms, legs or face, is not dangerous, but visceral (mid-body) fat is. The latter generates inflammation, and inflammation is a process that can shorten your life. It's complicated, but if you have inflammation, then your cells and tissues are attacked by the mediators of inflammation, and proteins are modified by it, too."

One of the overwhelming conditions of recent city life is the level of sedentariness it both inculcates and requires. The early adopters of urban living would also have been sedentary compared to their predecessors (we know this from their bone-density scans). What role might sedentariness play in terms of longevity?

"Sedentariness is considered universally bad for longevity. Usually, the consequences of a sedentary life get in the way of longevity: diabetes and metabolic disease – these are the real killers most likely to shorten life."

An example might be taken from the research Pes is conducting in Seulo, another of the island's blue zones. He tells me about a number of women who were tailors or seamstresses there. These are women who have spent their working lives sitting for many hours at sewing machines. "This might be considered a sedentary life, but there is an important exception. When I measured the calf circumference of the women, I found they were larger and more muscular than average." The pedal-operated sewing machines (which I suppose would be called manual) meant that even though they had spent a life sitting still, they had still burned sufficient calories for them to have derived the longevity benefits of being active. His next project was "to measure the energy expenditure involved in working these sewing machines." An inevitable outcome to this next study will be to provide guidance as to how work spaces might be optimized for workers, so that no matter how sedentary a job might be, there may be solutions that will keep them healthier for longer.

In Sardinia, a likely by-product of this more active sedentary life might also be some reduction in stress. I wondered whether there were national or even regional characteristics which make these centenarians less inclined to suffer from stress. Were these people just very laid back?

In the standard psychological tests that trial subjects are currently given, the centenarians rarely report any stress whatsoever, but this is not likely to have always been the case. Some psychologists call this the paradox of ageing, that as the body fails, the general psychological outlook of the individual improves. After all, the centenarians Pes works with had fought in wars and negotiated some of the massive political, social and economic upheavals of the 20th century. They were sometimes poor and were driven to migrate to find better wages and working conditions. Pes does not believe that they were not stressed, more that their ability to cope was a little different. They were never desperate, and never felt hopeless. They were always able to imagine a way out of their predicaments.

I thought again about the inhabitants of the Isle of Luggnagg in *Gulliver's Travels*. What kind of life had they had from the age of about 50? I'm only 48, and already suffer constant (though low-level) pain in my back. What might it be like with *another* 30, 40 or 50 years of age tattooed into it? Gianni's great uncle was what is called a supercentenarian (110 years or older). A trial by a group of gerontologists based at Boston University reported that 10 percent of supercentenarians made it to the *last three months* of their lives without being troubled by major age-related diseases^.

I was left with the enduring impression that I hadn't got things quite the right way round. I realized that I don't want to know why these people live such long lives, but why the rest of us live such comparatively short ones. And after all, it is no coincidence that Lugalbanda from the ancient Sumerian King List, who reigned for 1,200 years, was the only one who was listed as "the shepherd". Just like the longest-living Sardinians, neither had fully embraced the life of the city and all had integrated exercise into and throughout their working day, every day.

Despite all the accoutrements and paraphernalia medical research has produced over the centuries, Sardinian longevity puts the lifespans of city dwellers in the shade. Despite our investing hundreds of billions in healthcare, there are regions in high-income nations (such as the UK and the US) where life expectancy is still as low as in the mid-60s. Even in parts of London (Tower Hamlets, for example) people can only expect 54 years of good health.^ Metropolitan life, at least the way we like to lead it, is still, after over 4,000 years, not yet compatible with a long life.

These early settlements and cities made a permanent impression on human bodies. Today we are obsessed with exercise – ranging from a determination to do none whatsoever, through those set on bagging their 10,000 steps, to people who attend several exercise classes daily. It turns out that this activity is as deeply rooted in urban life as the idea of work itself. Not only that, but work and exercise are intimately connected with one another and can only emerge from within the hierarchies that metropolitan dwellers shape for themselves. To see this story at its clearest, we need to jump forward in time to a new city state at the heart of a new empire: the Ancient Greeks.

EXERCISE – A BAROMETER OF URBAN INEQUALITY

Throughout the entire Agricultural Revolution and into the Metropolitan Revolution, the driver for change was often war, but it was also comfort. Where there's a desire for comfort, there's a drive toward leisure. With settlements – and later cities – rose the idea of land ownership and new social hierarchies. Within these emergent hierarchies, to be one of the elite was to enjoy leisure time in which one might be entertained by musicians, actors or bouts of wrestling. In hunter-gathering communities there had been very little time or attention given to passive modes of entertainment or leisure in which the individual played the role of consumer. Instead, much more time was spent in conversation and social exchange – basically, gossip. Instead of sitting in front of PlayStations and Xboxes, hunter-gatherer children played collaborative games that mimicked the activities they would perform when mature adults. They played at building shelters, war, climbing and hunting. Games in which a score was kept, or in which there was a winner were rare to the point of non-existence. Play and education in hunter-gatherer communities were the same thing.

As city states began to rise the leisure activities undertaken by the rich were much more recognizable to Anthropocene humans, who require others to do the imaginative work for us so that we may sit in passive contemplation. The Greeks were the first culture to worry about what bingeing on leisure does to the body. They were well aware that sitting eating grapes while someone strums a harp was both a bit boring after a while and really not good for the body.

Historically, the extent to which a community is encouraged to exercise is indicative of the levels of inequality among its working people. The emergence and presence of exercise is a cultural barometer for something having gone awry in the ecology of labour. It means that work has changed so fundamentally that extra play or work has to be added to a routine to

The gymnasium, where exercise and philosophy met.
Often in the centre of a city and strictly men only.

make it more healthy. For hunter-gatherers, work and play were practically indistinguishable and exercise was a faraway dream of the future.

We think of exercise as natural and part of the human condition, but its nascence and spread was similar to that of the chair: it may be omnipresent now, but it has a surprisingly short history and was only practised by certain sectors of society. Moreover, it was only once we started to sit down in chairs that we realized we really ought to get out of them.

As crop yields rose, so did settlements. The settlements became villages, towns, ports and eventually cities. As conurbations grew, so did the necessity of transport, sedentary work and that ever-present symbol of urban malaise: exercise.

Long after the grains had settled after the Agricultural Revolution, the idea of exercise emerged in Greek culture in response to changes in working habits and patterns, and to leisured inactivity. Those who had thētes (the peasantry of Athens) and slaves to do their work for them soon realized that without some physical labour, their health would swirl down the plughole. Exercise was distinct from games and play. While the outcomes might have been similar, exercise was specifically carried out so that one's health might be improved, whereas games were played for all kinds of reasons: entertainment and amusement or competition and sport.

Exercise appeared only in societies, especially cities, where there was a profound workload differential. The wealthy men of Ancient Greece, deprived of work and with little else to do, invented a new place called the gymnasium (from *gymnós,* the Greek for "naked"). This was an open space in the city where they could strip off and gambol about naked, competing in made-up challenges to keep one another fit for war. The palaestra was a little like a gymnasium, but was smaller and focused on sports such as wrestling and boxing. This is what the men got up to while the thētes tilled the fields and the wives stayed within the confines of the home.

The Greeks were not the only ones to exercise. Exercise appeared in Roman culture, too. Cicero, the Roman politician and lawyer, celebrated the fact that, "It is exercise alone that supports the spirits and keeps the mind in vigour."^ Pliny the Younger, a writer and also a Roman lawyer, explained, "It is remarkable how one's wits are sharpened by physical exercise."^ Like their Greek workout buddies, these men were equally privileged and wealthy.

In common with so many aspects of modern life, exercise is part of something like a culturally implanted memory. It feels very natural to us, but after the Greeks and Romans, it practically disappeared. It peeped from behind the curtain in the early 19th century where it was found in unusual places such as Jane Austen novels. It might sound odd, but the land-owning strata of the society that Austen wrote about in her novels was not wholly dissimilar to the aristocrats of Athens. In Austen, it was the idle rich that were most in need of exercise.

The predominance of exercise in contemporary culture has emerged within living memory. My grandparents, who lived to the age of 80, certainly had no need of it. Their lives provided them with all the opportunities to move that they needed.

But in 2017 a study was published in the UK which reported that 41 percent of people aged between 40 and 60 (more than 15 million people) walked at a brisk pace for less than ten minutes per month.^ When I first read this, I assumed that "brisk" had been interpreted as unreasonably fast; perhaps something close to a slow jogging pace, which would have made the figure understandable. But no, for the purposes of this study, "brisk" was interpreted as faster than a 20-minute mile.

The research was part of a wider push to encourage people to be more active. Public Health England called it Active 10, and it is a little like the five-a-day campaign for fruit and vegetables. It does not ask too much of people and there are marked health benefits from engaging with it. Original government guidelines for exercise were 150 minutes per week (5 sessions of 30 minutes each). But Active 10 lowers its expectations as quickly as a needy partner in a doomed relationship. *If you won't do 30, puh-leeez, just do ten!* Even those ten minutes can have a marked impact and could slow or prevent the early onset of heart disease, type 2 diabetes, dementia, some cancers and disabilities.^ Although the middle-aged are the worst group affected, the report suggests that more than a quarter of the population fit into this toxically inactive band.

While ten minutes is a world better than nothing, if you're too busy to integrate 30 minutes of activity into your day, something has gone seriously awry in your priorities. If your life does not permit you to move, as the months go by, you will shave years and years of good life from your time on Earth.

The push to get us to exercise is a sign of just how out of step we are in our lives. Sedentary workers earn while relying on others to grow and source their food and all we have to do is stroll to a shop, or order groceries online to have someone bring them to our door.

It is odd to think that the foundation of today's Olympics was the result of early athletes being forced to take up exercise because slaves and peasants did all their work for them. And, as we have seen from as far back as Uruk, running concurrently with humans' move into the city was their pursuit of longevity. The Greeks and the Romans understood that even though their slave class did their work for them, exercise and physical activity were essential for a long and sane life.

Today, those Sardinians for whom a cereal-based diet is still not an easy option are outliving their relatives who move to the cities. Even though they might share the same genes, the new metropolitan dwellers drive when they should walk; they grab an espresso and a pastry instead of some pecorino, grape juice, goat's milk and a little bread. Their diet may be more varied than those who live in villages in the mountains, but more choice is not the same as having the best options from which to choose. Even though the residents of the less accessible villages are sustained by a lifestyle that is agricultural in origin, it has not yet been dominated by the monoculture of refined carbohydrates that powers our cities. And the way they live means that they are active to a point at which they have no need of something as absurd as exercise.

Are exercise and inequality uniquely linked? This is one of those questions that it is best not to inspect too closely, but the short answer is "yes". The question of whether exercise and the way we undertake it is good for us will be deferred until we reach the 20th century, but Gianni Pes's work on extreme longevity in populations which have no access to a local gym, workout classes, park runs, cycling clubs, Pilates and do not have a cross-trainer, rowing machine or treadmill gathering dust in their spare room is more than a clue for now.

WINDING BACK

The principal cases in this chapter all share genetic elements: the jaw, stunted height (it applies to myopia, too), but they are also limited because these are developmental problems, and once adulthood is crested they may not be edited in later life. Once your height is set, you will not grow even if your diet improves in your mid-twenties. Once the shape of your jaw is established, it is just that, established – although bone density can always dip and rise. Still, there are easy things you can be do to restore oral pH or, you know, add a few years to your life.

1 Consume more vitamin D and "vitamin N"

Regular access to daylight is important for all of us, but particularly for children. While spending hours outside in the sun carries obvious risks of over-exposure, it also provides a cascade of associated benefits including movement and the production of vitamin D, as well as a whole raft of nature-derived psychological benefits that are so essential they should be called vitamin N. With iPads, iPhones and laptops now increasingly able to produce more than 500 nits of brightness (that is, their screens are bright enough to be used in direct sunlight), and with improved water resistance, they are more amenable to outdoor use (the moped gangs can't wait).

2 The mandible is for chewing

Encouraging the consumption of chewy foods, particularly by the young, can aid jaw development and saliva production, almost certainly lowering the likelihood of malocclusions, but can also be beneficial for dental health. Chewing gum (foul stuff that I cannot possibly recommend) is good for this, and it is also useful for restoring the pH in the mouth after eating (a glass of water is just as good, but you don't get the mandible workout).

3 And chew some more

It is never too late to stimulate mandible bone density, which may encourage root health in your teeth, too. Just as with any muscle and bone in your body, working the jaw will stimulate growth. I'm not recommending eating raw meat, but there are lots of foodstuffs that are healthy and will give you a good mandible workout: celery, carrots, kohlrabi, nuts and seeds (the latter may be calorific, but if they are not processed we can only absorb about two-thirds of their fat and they have been shown to suppress appetite).

4 Brush before eating, not after

The chemical and physical abrasion of brushing while the pH of the mouth is acidic damages a greater surface area of the enamel on the teeth. It is probably worse to brush after a meal than not at all. Better to do it before eating, or try recommendation 5.

5 Drink water

Consuming a glass of water or eating something that restores the pH of your mouth (such as a chunk of cheese or some yogurt), straight after a meal shortens the time frame during which acids attack the enamel. Eating fewer meals and no snacks will help, too.

6 Weights are the worst

During my interview with Gianni Pes he explained, "The right kind of exercise for longevity is aerobic. Weights are the worst." But it is tricky to set hard and fast rules here. Resistance training, though proven to have fewer metabolic and psychological advantages than other kinds of activity, is very good at keeping core strength in muscles that might weaken with age. Maintaining power in the muscles is important for living an independent life. So resistance training won't add years to your life, but maintaining strength will add more years of independence. If your resistance training is already taken care of (by a moderately active job, for example), then Pes is right. Aerobic exercise, spread over long periods at low intensity, is proven to be best for longevity and also makes it much more statistically likely that you will not have to be admitted to a nursing home during retirement.

7 Choose the right exercise

Going back to the laws that govern the strength of hard and soft tissues (the move it or lose it ones), then problems are likely to arise when you combine the growth in the mean weight of a population with a diminution in the amount of time spent on their feet. Once the feet and ankles cannot cope, neither can the knees. The sufferer could just lose a little weight (if that is their particular problem), but it is better to try to address the weakness that has probably crept up the legs from the feet like rising mould. One benefit weights have is to stimulate bone density, but other forms of exercise do this too. Not all exercises were created equal, though. Cycling, swimming and other non-weight bearing activities (though great for many exercise-related rewards) do not stimulate the osteoclasts to go to work repairing and thickening bone. But a study in 2018 showed that cyclists who remained very active and rode bikes into their 70s and 80s had the immune systems of people in their 20s.^

8 Good news about booze

Many researchers interested in longevity see red wine as one of the key ingredients to a long life. Roger Corder, Professor of Experimental Therapeutics at the William Harvey Research Institute at Queen Mary University, London, and author of *The Wine Diet*, has explained to our relief that red wine can function as a miracle cure for heart disease, some cancers and even erectile dysfunction.^ The bad news is that alcohol has recently been linked to DNA damage in stem cells.

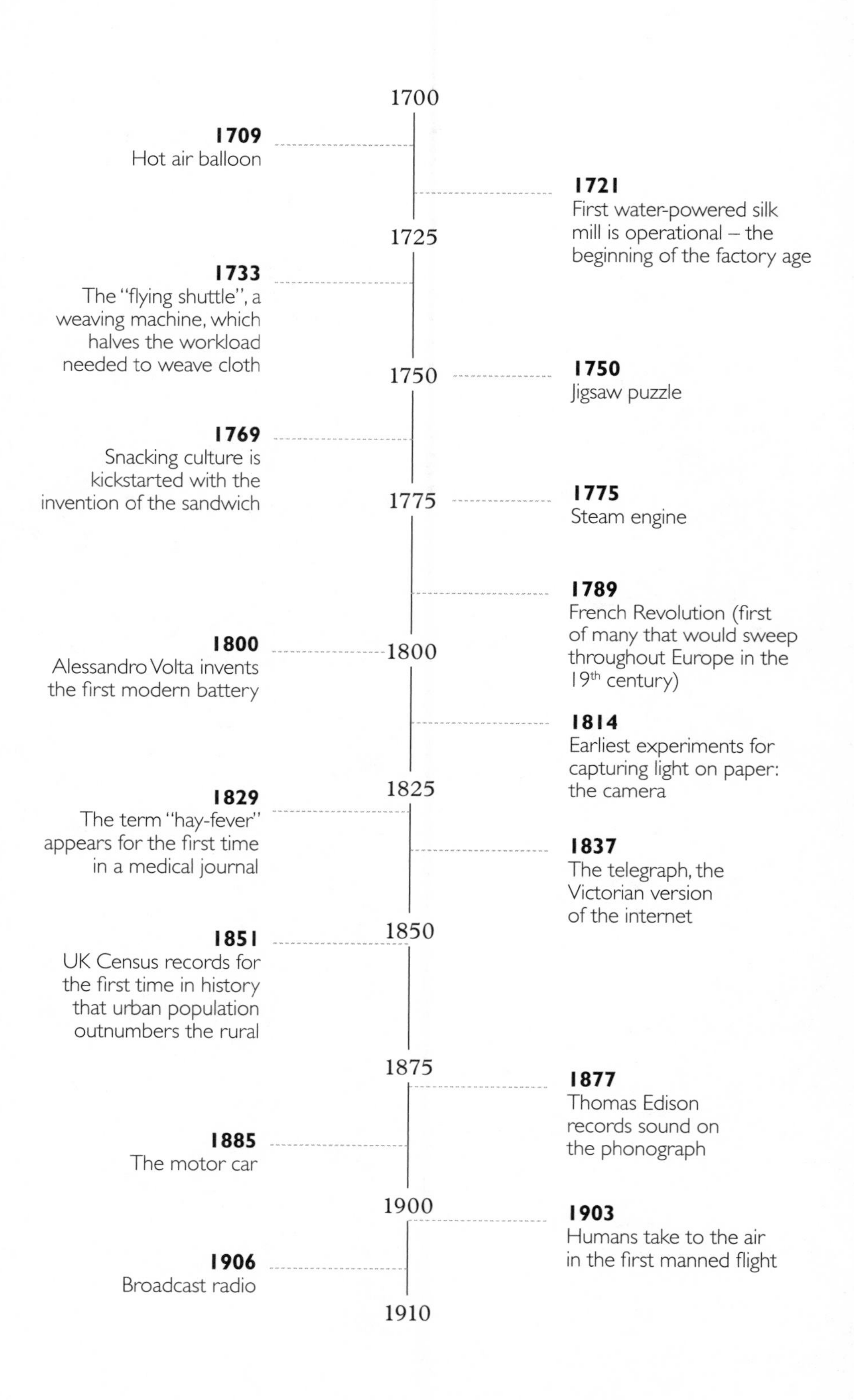
1700
1709
Hot air balloon
1721
First water-powered silk mill is operational – the beginning of the factory age
1725
1733
The "flying shuttle", a weaving machine, which halves the workload needed to weave cloth
1750
1750
Jigsaw puzzle
1769
Snacking culture is kickstarted with the invention of the sandwich
1775
1775
Steam engine
1789
French Revolution (first of many that would sweep throughout Europe in the 19th century)
1800
1800
Alessandro Volta invents the first modern battery
1814
Earliest experiments for capturing light on paper: the camera
1825
1829
The term "hay-fever" appears for the first time in a medical journal
1837
The telegraph, the Victorian version of the internet
1850
1851
UK Census records for the first time in history that urban population outnumbers the rural
1875
1877
Thomas Edison records sound on the phonograph
1885
The motor car
1900
1903
Humans take to the air in the first manned flight
1906
Broadcast radio
1910

Part III

1700–1910

Frontispiece to
E W Duffin,
On Deformities of the Spine (1848)

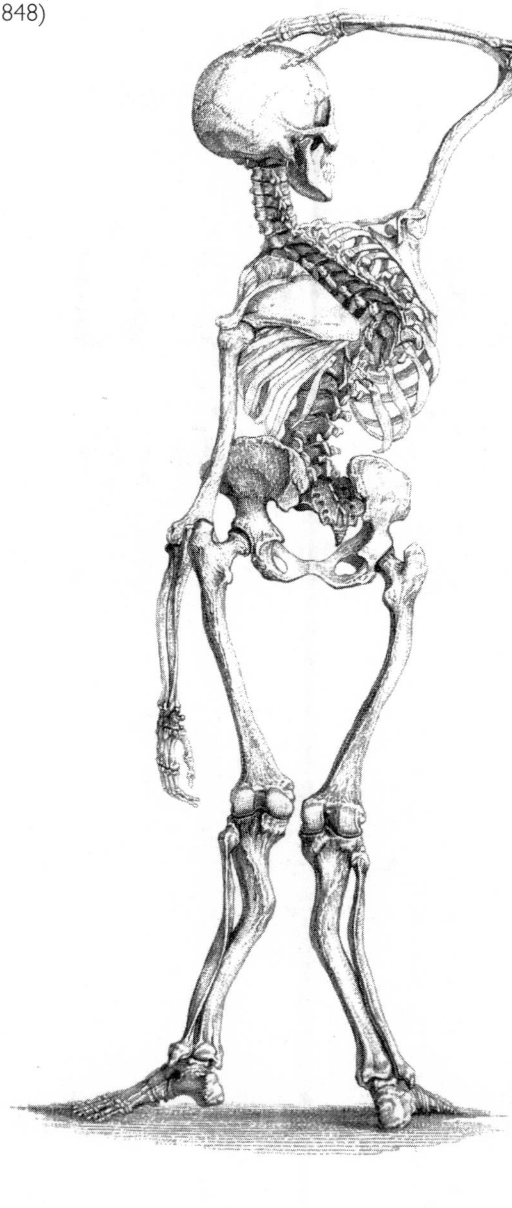

MINES, SPINES, SMOKE & STEAM

> You must either make a tool of the creature, or a man of him. You cannot make both. Men were not intended to work with the accuracy of tools, to be precise and perfect in all their actions. If you will have that precision out of them, and make their fingers measure degrees like cog-wheels, and their arms strike curves like compasses, you must unhumanize them.
>
> John Ruskin, *The Stones of Venice* (1853)

For thousands of years agriculture coexisted comfortably with the rising metropolises. Cities grew in size, frequency and sophistication. Numerous inventions made life for farmers and townsfolk either easier or more efficient. Change was slow: an agriculturalist from Sumer in 2500BCE would, with a shiver at the cold, recognize the practices of a northern European peasant working the land while swearing fealty to his or her lord and struggling to make the rent. Take that Sumerian farmhand further forward in time to the early 19th century, and they would have no idea how to handle a spinning jenny or a screw-cutting lathe.[1]

Shunt the farmhand forward a couple more centuries and show them the 22.5 million acres (about 35,000 square miles) that comprises the Mudanjiang City Mega Farm in Heilongjiang, China; would these be recognizable to them? Could our Sumerian make sense of a factory farm the same size as Portugal?

Change accelerated when humans crested the Industrial Revolution in

1 The spinning jenny revolutionized spinning cotton with its multi-spindle preparation of threads for weaving; it was a key invention. The screw-cutting lathe made it possible for screws to go into mass-production because of its ability to create accurate threads on screws for wide applications.

the 18th century. And as the landscape and environment changed at a great pace, so did their bodies.

So much of who we are now, and of how we look, starts in the period of the Industrial Revolution as those ecologies of labour shifted to become increasingly specialized. It is hard to overestimate the revolution's influence on our ways of thinking, our politics, our diets, our need for comfort, our relationship with machines, with the environment, with technology and our bodies.

With those seeds of new behaviour planted long ago, new bodies began to emerge, shaped by the kinds of work that those bodies were asked, or being made, to do. Bodies became mangled by labour so squalid and offensive to our biomechanics as to permanently leave its mark upon them. The story of the human body is there in the documents and data from which traditional history usually draws. But it is there in fiction, too, informing us as it does about the assumptions, not only of those who penned the stories, but of the times and places in which those stories were read. When an entire genre of fiction emerges from the changes wrought by new modes of work you know the situation is serious.

The industrial and social problem novels of the 1830s, 1840s and 1850s were quite short on solutions to the problems of industrial capitalism, although they were full of bile and outrage at the absurd and disgusting inequalities that were emerging in an ever-widening rift between rich and poor. The novels are clogged with tales of work-related injury and disability, which meant that bodies (and their disabilities) inevitably took on connotations of class.

Working-class autobiographers of the 19th century such as William Dodd and Robert Blincoe reported the explosion of disability among factory workers.^ Bodies shrank as the diet of the poorest in the new urban centres became ever simpler and less nutritious. Other bodies grew, feeding off the labour and production of those capitalists who benefited from the profits generated by new machinery. But it is not just these figures that tell the story of the industrial body.

During the 19th century levels of literacy were changing rapidly; more writers meant more readers and the proliferation of storytelling and journalism meant that there was a wealth of information about what was happening to the bodies of those who worked – and those who need not. To see how different things were (both in terms of data collection and as a snapshot of the severity of working and living conditions in the 19th century) let's look again at human bodies as an economic gauge or symbol.

ECONOMIC BODIES

After several thousand years of advances in farming techniques and agriculture, the European male was on average just 2.5cm (1in) taller than his ancient ancestors had been. A safe inference from the data might be that from that point onward, humans started to grow gradually toward our current mean height (in Europe) of about 178cm (5ft 10in). This would be wrong. That extra height was soon lost when people flooded into the factories, when the bodies of the working classes shrank faster than a cashmere pullover in a hot wash.

In 1837 Charles Wing, surgeon to the Metropolitan Hospital for Children, reported the average height of 13-year-old factory workers as 133cm (4ft 4½in).^ A mass observation by surgeons from 1836–37 reported that 14-year-olds averaged 140cm (4ft 7½in) – an inch shorter than their counterparts who lived in rural settings.^ James Harrison, a surgeon in Preston, found a mean height of 152cm (5ft) in 159 of the 17- to 18-year-olds he measured in 1836.^ Harrison explained his cautiousness in gathering such data by stating that these measurements were taken "without clogs". The average fifteen-year-old male in Britain today is a giant by comparison with these workers: 173cm (5ft 8in).

The work of many of these bodies had been both simplified and extended so far that their elastic broke. As workers poured from villages and small towns into the new urban centres, their work took on an increasing homogeneity.

Machines require operation by the body but less movement of it. The miles earlier humans had walked and run on the savannah would have shrunk dramatically once workers entered the factories. Until this point, about 80–85 percent of the people in the world would have been engaged in farming of some kind and roughly the same percentage would have lived on the land provided for them. After the end of the 19th century, the percentage of people farming dropped to about 1 percent. For many thousands of years, the top speed reached by any human was about 27km/h (17mph) on horseback.

Instead of working in the fields, labourers now stood for extraordinarily long periods. Their diet, if you can call it that, was about as nutritious as eating an old piece of newspaper. Poor nutrition and the dehydration that resulted from having only occasional access to clean water had catastrophic long-term effects on human health. Nutrition is important for microcirculation, which particularly affects joint health and repair, as does dehydration. The pressure of standing squeezes water from intervertebral discs and joint cartilage alike – both can lead to permanent degeneration in their structures and consequent morbidity in the worker.

The workers who left their villages for the cities in the mid-19th century were time travellers. They escaped their feudal cottages and walked 16km (10 miles) down the road to find themselves in a new, modern era of human development: the coal-coloured capitalist city. To find their way all they needed to do was head for the towering plumes of smoke.

As long ago as the time of the Ancient Greeks and Egyptians, the tanned body was gendered (we know this from their art). The division between public and private was written visibly on the body. Women should be pale-skinned (rich enough to stay sheltered indoors), their realm was the home; the men were dark-skinned as they were at war, wrestling in the open-air gymnasium or working the land.

By the time of the Tudors, and most famously that tundra-faced monarch Elizabeth I – the snow queen herself – skin colour became more firmly economic. The paler you were, the richer you were likelier to be. The peasantry forever out tilling the fields would have been as brown as berries, so the aristocratic fashion became the opposite. And so the aristocracy powdered themselves with poisonous compounds rather than be mistaken for someone who had publicly demeaned themselves by going out in the sun.

For whitening, Elizabeth I used Venetian ceruse, a thick lead paste. Lead was essential to the mixture as it gave the compound a smooth and even whiteness, so it looked more natural and spoke of a kind of racial purity appropriate to one whom God himself had placed on the throne of England – at least, the Elizabethans believed it did. Skin tone was economic.

In China, the tradition of foot binding, which lasted for well over a thousand years, was similarly economic. The ideal of femininity, the 10cm (4in) foot, was tied up with ideas about the erotic appeal of the "lotus gait", with its small and swaying shifting steps. But it began among the wealthiest families and boasted of the host's freedom from the need to move any significant distance. Chinese women did not need their feet to work (in both senses of the word).

The 19th-century body was equally economic. Think of the importance of composure, demeanour and posture in Jane Austen's books, or the fictional school of deportment in Dickens' *Bleak House* (1852) and it begins to seem that the ability to stand up straight became a class issue.

In this world, the stooped and the mangled had debased themselves with the kind of work that had stooped and mangled their bodies. This was a culture in which being able to stand up straight was as political and social a statement as it was an embodied one.

In countless industrial novels from the period, workers are described as being pale, thin, grey in pallor and hunched. Their work had made their flesh; their bodies were economic signals to those around them. And, just as the fashion in the Tudor period was one that differentiated the rich from the poor, so too was it in Victorian culture.

In a world where calories were difficult both to acquire and hold on to, the perfect male body in the Victorian period was one that boasted of a little excess. The artist Nickolay Lamm recently studied photos and illustrations of the male body from the last 150 years, concluding that in the 1870s a little extra weight was a sign of burgeoning health.^ This ideal morphed between the wars into a man who looked more like a healthy, athletic manual worker, evenly built with a strong posture, rather than that of the Victorian gentleman who was beginning to look a little fat.[2]

At all points in history, these bodily ideals were economic, in so far as they rarely reflected the bodies of the working classes, but more often those acquired by sheer effort, usually by people with sufficient time and money to invest in them. Victorian factory hands had no such time and so lost the ability to stand up straight.

Those who worked machines in the mills of the 18th and 19th centuries were often required to stand in one place repeating movements in a manner not unlike the machines they were operating.

Before Victoria's reign had begun, anxiety about what was happening to the bodies of England's workforce was already widespread. In 1833, Peter Gaskell

---

2 This slim look predominated until the counter-cultural shift of the 1960s and 1970s, when slim became positively skinny. The new "rockstar" body was one that clearly differentiated men from their fathers, or bodies in the corporate mainstream. It broadcast a move away from active physical pursuits into the psychosocial pleasures of getting baked, blazed or blitzed. In the 1980s everything became bigger, tougher and more robust. Women's fashion went for bold colours, sharp edges and shoulder pads. For men, the body-builder look went mainstream on cinema and TV, with role models such as Arnold Schwarzenegger, Sylvester Stallone or Dolph Lundgren. Today we are somewhere between the two, and a muscular but athletic body is widely popular.

produced a study which considered the lot of the pre-industrial worker, praising such "employment of a healthy nature" (which, comparatively, it was).^ Reading it is like listening to a mournful song by The Smiths, played at half speed. Gaskell liked the sense of social cohesion that came with working from home because it helped to keep families and people together in tidy domestic units.

He despaired of the long working days, not only for the sake of the workers, but because their working patterns destroyed "social and domestic relations". His book goes on to look at all aspects of workers' lives: what happens during their working day, their living conditions, their diseases, illnesses and pathologies. And while he was able to look at some of the great improvements that industrialization brought about, the study leaves us in no doubt about the appalling human cost:

> ... an uglier set of men and women, of boys and girls, taking them in the mass, it would be impossible to congregate in a smaller compass. Their complexion is sallow... Their stature low... Their limbs slender... Great numbers of girls and women walking lamely or awkwardly, with raised chests and spinal flexures. Nearly all have flat feet, accompanied with a down-tread, differing very widely from the elasticity of action in the foot and ankle, attendant upon perfect formation.

Flat-footedness should have been exceptionally rare in any population, but for these workers it had become the norm. Human feet adapted and developed over millions of years to allow us to walk long distances, and sometimes run them, too. The foot's physiology is all about motion and once feet are starved of the oxygen of movement, they go into decline.

The osteoblasts in our bones go about their work of thinning its density. Because the feet are not covering the daily distance they were built for, the body makes the sensible decision that the bones will be better adapted if they are lighter (the denser the bone, the more calories required to move it). The muscles in the foot also atrophy.

Our bones and muscles work on the basis of "move it or lose it". If you stand around all day working a machine and deprived of sunlight, you are guaranteed to "lose it".

The intrinsic muscles – four layers of arch muscles – gradually become thinner until they are no longer capable of supporting the structure as intended. There is a core level of bone density and musculature beyond which the body will not atrophy, but wherever that point may be, we can see from this evidence (and the epidemic of flat feet today) that our feet flatten and fail before we reach it.

Throughout history and prehistory billions of humans will have possessed a genetic tendency toward flat-footedness. But when humans roamed 8–14km (5–9 miles) each and every day, the tendency rarely, if ever, had an opportunity to express itself. Despite our genetic intentions, we know that genes only ever create a propensity toward a particular outcome. They also need the right environment or triggers, and this they found in the pathological working practices of the 19th century, just as they do now.

Until the Industrial Revolution, the human foot was constantly stimulated by activity and remained strong; bones remained dense and the soft tissues were tensile and sturdy, so the arch of the foot would have remained high. Two of the muscles (the *flexor hallucis longus* and the *flexor digitorum longus*) attach toward the back of the knee and run down the calf, around the hinge of the heel and all the way down to the tips of the toes. These muscles pull the toes medially back toward you to help create that lift in the arch.

Like our organs, once one part of a limb starts to fail, another may try to take over. When the arch is no longer playing its part in supporting the frame, the ankle tries, and the knee, and the hips, and the spine and the neck try to help too, but they can't.

Humans never fully adapted to life on the ground – not as completely as horses, sheep or goats. (Or is it the case that we have not yet adapted to sedentary life on the ground?)

The mechanism early humans developed was complex, powerful, ingenious – even beautiful – but without the right kind of sustenance and stimulation, it is ultimately fragile. In order for our feet and our bodies to stay healthy, they need movement in the way that stomachs need food and skin needs sunlight. Once our feet are weak, the weakness can spread to other parts of the human body like a virus.

Victorian factory workers suffered from a mixture of overwork and malnutrition and this, coupled with a new kind of upright sedentary work, created a perfect storm for pathology to run as wild as Luddite machine-breakers on a rampage.

How a flat foot can affect biomechanics throughout the body.

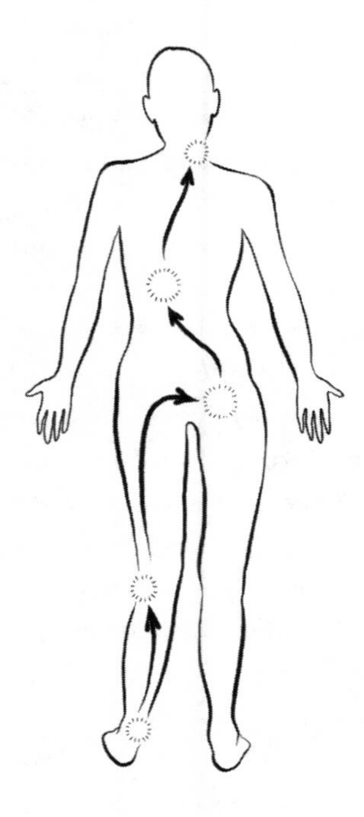

Knock-kneed people do not exist in pre-industrial fiction, but by the time we crest the 1830s, they are a common sight and habitually shuffle into view. In Dickens's *Oliver Twist* one of the younger characters (Noah Claypole) is described as having a body unable to support itself, like an eager shoot spreading its leaves widely to welcome the sun, only to find that the stem can no longer support itself. He is "one of those long-limbed, knock-kneed, shambling, bony people".^What we have here is some good old-fashioned biological determinism.

In the late Victorian period, there was an obsession with physiognomy, and the notion that certain kinds of criminality could be read upon the face of the criminal.[3] A little earlier there was phrenology. This was the reading of the crests and dips on a person's skull, and the belief that cranial bumps could provide a clue to one's true biological tendencies. The idea was that with this knowledge an individual could address their natural amativeness (propensity for love or sexual feelings), for example.

But there is also sociological determinism – the idea that the way we live

3 Pointy noses and small eyes did not do too well here.

our lives also determines the shape of our bodies. In his book *The Conditions of the Working Class in England*, Friedrich Engels observed in the chapter on factory hands, “The knees were bent inward, the ligaments very often relaxed and enfeebled”.^ Much later, observing miners he noticed, “distortions of the legs, knees bent inwards and feet bent outwards ... and they are so frequent that in Yorkshire and Lancashire, as in Northumberland and Durham, the assertion is made by witnesses, not only by physicians, that a miner may be recognised by his shape among a hundred persons”: sociological determinism.

Peter Gaskell (*see* page 133) eyed the bony bodies of working-class adults and children and found a “very general bowing of the legs” and children “many with limbs bent” and riddled with rickets, once called “the English Disease”.

What happened at the beginning of the 19th century? Why were the fictional and factional bodies of the working class collapsing?

Legs that give way under their weight are fairly common in younger children up to about the age of six, and are a quite normal part of development, with the legs gradually straightening again before puberty.

During intense growth periods, between the ages of one and five and during puberty, for example, new bone is made on the growth plates (toward the ends of bones, especially the femurs) and it takes essential ingredients from the blood to do so. If phosphate and calcium are not available at a time when the bone is ready grow, and vitamin D is not there to help the nutrients on their way, the bone cannot form correctly. Instead of being dense and firm, the new osseous tissue is frayed and weak, which leads to distinctive bowing (of the legs) and probably to swelling, especially at the ankle and knee joints.

Without treatment, the bowing becomes permanent. We need sunlight to make vitamin D (we cannot absorb sufficient amounts from diet alone) and at northerly latitudes, between November and March, there is not a great deal of sunshine to be had. One of vitamin D’s functions is to aid the absorption of calcium, which is equally important; one without the other is of little use. (The spread of rickets among Victorian child labourers was evidence that these children were not only poorly fed, but that they rarely saw daylight either.)

Miners’ shortage of sunlight (and that of many others like them) was exacerbated by spending their nights toiling for coal, and their days sleeping in recovery. Knock knees in this case became the very essence of the

Anthropocene body, with biomechanical symptoms and problems derived from extracting fossil fuels. The modes of extraction used to access the fuel meant that the bodies of the men and women who worked in the mines were destroyed by the labour needed to fire the furnaces of the Industrial Revolution.

Now when adults develop knock knees the cause can be a number of problems: an injury or an infection that has attacked the leg or knees, or underlying genetic conditions. The most likely cause, though, is excessive wear and pressure.

INDUSTRIAL MANGE; SKELETAL COMPLAINTS

Lack of access to sunlight, a diet deficient in vitamin D and/or calcium, and overwork all might have played a part in mangling the bodies of the 19th-century workforce, but new work patterns gathered all these elements together as neatly as finely-woven lace. And the problems caused were not only skeletal.

There are more than 200 kinds of arthritis, the most common types being rheumatoid and osteoarthritis.[4] The causes of osteoarthritis are still largely unknown, but it is now the most common joint disease and source of chronic disability in the West. This condition can attack many joints, although the knee is both one of the most common and the most important. Like a gateway drug that draws you in ever deeper, osteoarthritis in the knees makes any sort of weight-bearing movement exceptionally painful and it now affects tens of millions of us.

Osteoarthritis is a condition in which the cartilage that protects the bones in a joint wears away. Cartilage is like the toughest, smoothest and most perfect hard rubber you can imagine, and it is wrapped around the ends of our bones to help ease and lubricate their movement. In human embryos, the skeleton is made entirely of cartilage before ossification, at which point only a little of the substance is left (our ears, noses and throats are made of it). Unlike other connective tissues, cartilage has no nerve endings, so it is hard to work out when damage might be done to it, and because it has no blood

4 Arthritis, from the Greek *arthron*, for joint, has been around a long time, dating back to the 15th century.

supply it can be very slow to heal.

In an arthritic joint, the icy-smooth cartilage has been worn away and inferior cartilage might have been created in its stead, with water taking up any empty space. The ability to absorb shock in the joint is severely diminished and the surrounding structures (tendons, ligaments, bones) are recruited to help deal with the load, leading to further damage and inflammation because they are not up to the job. Instead, the bones grind against one another, leaving permanent scratch marks on them called "eburnation".

This is Anthropocene body pain. A 2016 study estimated that of the 1.25 billion people in India up to 28.7 percent suffered from knee osteoarthritis.^ In the UK, according to figures published in 2013, "around a third of people aged 45 years and over (a total of 8.75 million) have sought treatment for osteoarthritis".^ In the US, the figures are higher, with 30 million adults affected. In this 2017 research a team at Harvard reported their first findings from a much larger study. They analyzed the prevalence of the condition in the remains of those over the age of 50 at their death.^ These samples ranged from the beginning of the 19th century through to the modern post-industrial era, in which the BMI at death of the cadaverous remains had been recorded (a total of 2,400 skeletons in all). The team, led by Ian J Wallace, also studied 176 skeletons of early hunter-gatherers and farmers (estimated again at 50 years old+) from between 4000BCE through to the early 18th century. For years it has been assumed that osteoarthritis is a disease of ageing, on the increase because our population is getting older. It has also been connected to BMI, and the idea that overloading a joint with extra body weight damages it. But the team's results question both assumptions.

In the pre-historic samples (and because eburnation is so easily recognized on a skeleton) only eight percent of remains studied in the fossil record showed evidence of the disease. In the early industrial samples (those taken from the 19th century), the prevalence actually dropped to six percent. But in the post-industrial era, particularly from the Second World War onward, pervasiveness doubled. In these remains, the condition was found at a rate of 16 percent. The results also suggest that there is insufficient causal evidence to indicate a link between the condition as simple as age and BMI.

Is the drop from eight percent (the levels of osteoarthritis in prehistory) to six percent (in the 1800s) statistically significant? Did the level fall because most work was undertaken while standing? With the sample sizes, it is hard to say. But the Victorians were more mobile than we are today. With public transport in its infancy during the mid-Victorian period, London clerks who lived in the suburbs would have had to walk many miles to and from their offices. So many did so, in fact, that the granite slabs of London Bridge were worn smooth and had to be roughened with mallets and chisels. There is a lot of noise in the numbers.

The jump from six percent to sixteen percent in a century is statistically significant. As with smoking and cancer, the scientifically-confirmed links of specific behaviour that lead to osteoarthritis are still not known. There could be many of them; the most likely I think is our activity levels.

It is almost absurd that a condition as minor as a sore knee could be the beginning of terminal health complications, but this makes sense, because like a number of other conditions, osteoarthritis is a gateway pathology. It can be the first of several phases of disability that escalate and mature as new pathologies express themselves in the soils of those that preceded them.

It's funny to think that for thousands of years we have been hearing about Gilgamesh and his search for the secret of more life. He found it on a quest, when all along it was his ability to stay mobile, to walk while searching, that was endowing him with more life.

The journey itself was the elixir. The quest complemented the search for longevity. Long life is not to be found in a special plant by the river Tigris, but in strong knees – knees made strong because they have been used.

Ironically, it is in the 19th century, the very period that saw a dip in osteoarthritis, that we find the beginnings of its expansion and spread in the next generation.

TEACH US TO SIT STILL

It all starts with a metaphorical seed planted some time between the Agricultural and Metropolitan Revolutions. In early cities, the concept of leisure emerged out of the entropy created by land ownership and social inequality. From its very earliest representations, the essential ingredient of leisure seemed to be that it must be enjoyed while reclining... in a chair. And as the Industrial Revolution wore on, new working patterns emerged that also required stillness (tailoring, lace work, accountancy and administration). There were hundreds of such occupations, but perhaps the greatest contributor to emerging sedentary life was the way we trained and recruited our children into it, too.

Chairs seem as old as time itself. I think most people would assume their invention to be Palaeolithic, but this omnipresent prosthetic for the modern human body has a surprisingly short history. While there were chairs in early city culture, the seed grew slowly and had to wait for the ideal environment before it could germinate. That environment was the 19th century, where it spread quickly. In the last 200 years the chair has become one of the most potent symbols of the Anthropocene body. How we ended up in a world that now has billions more chairs than people is a story grounded in the Victorian period.

Do a quick count of the chairs in your house. At a glance, mine has four in the living room, another five in the kitchen and one in the study. Ten chairs! No, wait; in the garden there are two more chairs, and a couple of two-seater benches, too. So that's 16. Should I add the seats in my car? That's another four. Gathering up a couple of fold-away chairs for emergencies brings the total to 22. There are only two of us.

Now do a quick tally of all the chairs you might have sat in today: your office, your colleague's office, the place you ate your lunch, the meeting room, the tube, the car, the train, your dining table and so on. When we factor in the many thousands in stadia and concert halls, cinemas, doctor's surgeries, hospitals, theatres, the millions in office blocks around the world, in cafés, pubs, bars, clubs, restaurants and churches as well as in all the universities and colleges with their lecture theatres and classrooms, then a conservative estimate of the number of chairs in the world could not possibly be lower than at least seven per person. Applying that logic then there is approximately 52.5 billion of them. Surely chairs should be one of the universal signals of the arrival of the Anthropocene? They are to be found on every continent,

A typical Klismos chair (left) – with a rounded back and tapering, outcurved legs; (right) a plastic one-piece mould injection chair in an Anthropocene setting.

and there will be several million lying discarded on seabeds.

There are some very old representations of chairs, such as a clay model of one from about 6,000 years ago and those depicted in Greek and Mesopotamian carvings. A few Egyptian chairs still exist, and they are clearly symbols of great wealth and power. The chair was simplified a little by the Greeks, who produced the stunningly stylish klismos chair in the 5th century BCE (none as beautifully simple appear for another 2,300 years).

The Aztecs had chairs, too, but they were for rulers and dignitaries. They began to appear throughout Europe a few centuries into the common era. There's a 6th-century chair in the cathedral at Ravenna, and others began to appear in Chinese, Japanese, Korean and Turkish cultures around the same period. A few centuries later they popped up in Glastonbury in Somerset and in early Franco-German Merovingian culture.

Chairs had been around for thousands of years at this point, yet because of their persistent association with power and wealth, they were about as widely used by the peasantry as a crown.

While they began to appear with a little more frequency in the early modern period, it seems that they became much more widely popular in the 18th and 19th centuries during the Industrial Revolution.

The role of the chair in literature is particularly illustrative. There are no chairs in Homer's *Iliad* or the *Odyssey*; the latter only has benches for the 108 suitors trying to steal Odysseus's wife, Penelope. Neither are any chairs mentioned in the Bible; but Shakespeare's *King Lear* mentions them three times (not surprisingly given the royal prefix of its title). *Hamlet*, in contrast, has zero. By the time we swing by that previously discussed tipping point in human history (1851) Dickens's *Bleak House* of a year or so later is a veritable warehouse full of them where they are mentioned 187 times.

Did supply suddenly meet demand in the mid-19th century? The answer is to do with power and modes of production. The French Revolution struck Europe as though it was the epicentre of a worldquake. Other forms of social and economic arrangement, such as the feudal system, did not trade on the fact that they were meritocratic (if you work hard, you will be rewarded with a better position in society). In the feudal system, there was no social mobility; if you were born a peasant, you would live as a peasant, work hard as a peasant and die as one, too. If you were a peasant or a noble then you had to live and dress and behave as a peasant or a noble.

Only when a populace is imbued with the idea that things might be different, that there are other ways to organize society, that they themselves might be different, can revolutions happen and things begin to change. In preceding social systems, one's place in society was locked, natural and internalized, but once that was torn down then everything was up for grabs. We can sell our labour to whoever we wish, supposedly. We can wear whatever we want, supposedly. And, we can sit down on anything we like, ideally with armrests and a soft cushion. Once the chair's symbolic potential as a signifier of social hierarchy was removed, we were all free to have them.

A simple chair was relatively easily come by, but upholstered chairs were prohibitively expensive. The fashion for the new reclining culture (imported from the French court of the 18th century) spread into the pages of 18th- and 19th-century fiction.

To accentuate the cultural permeation of upholstery in the period, William Cowper, a widely read poet in his day, published in 1785, *The Task: a poem, in six books*. The first of these books was entitled *The Sofa.*^ The sofa had once been the lofty inspiration for philosophical meditation and was used to satirize the enervation of the idle and irresponsibly wealthy, but now, a couple of centuries later, the sofa is the cornerstone of every modern living room.

With the Industrial Revolution in full sway and the discovery of new materials, the possibilities for the mass production of chairs took over. Their price generally plummeted and they were widely adopted – hence their sudden prevalence in 19th-century fiction.

When one-piece injection moulded plastic chairs first appeared in 1960, their stage was set for world domination.

Today, with more than 50 billion of them on the planet, they can be snapped up on auction sites for 1p, or on sites such as Craigslist or Freecycle for even less than that. In the modern world, where so much has changed since those early settlements, the parallels are as curious as they are quiet. We still live and work in strict hierarchies and social mobility remains out of reach for most; we have to exercise because others provide our food for us, and the boss still gets the most luxurious and ergonomic chair in the office.

THE EDUCATION FACTORY

With the exponential growth in the use and abuse of chairs in the 19th century also began the very steep increase in the range and type of sedentary behaviours. It would require the gathering of several other factors before it became possible for us to sit for the 15 hours a day that many of us manage now; but here, at the height of the economic dominance of the Victorian empire, is the next chapter in the story of the chair.

Where hunter-gatherers learned through imitation and play, the bold young women and men at the vanguard of imperial expansion – the next generation of Victorians – were taught about their domain using different methods. At a glance, these ways of teaching seem completely bizarre and unrelated to any real world application. The students' learning environment was not so much a preparation for the world of work and commerce but was overwhelmingly focused on the inculcation of obedience. It was during this period, and by looking at the Anthropocene body in training, that we can see new habits becoming firmly established, and in the space of a generation or so, these became "traditional".

For many centuries, humans had access to chairs but did not use them. Sitting is something that we have had to learn to do, and the role, not only of schooling, but of the ideas that lie at the core of our educational philosophies, are also in the frame for the demise of our gluteal muscles since the time humans first sprang from the savannahs.

Formalized education had been around at least since the Sumerians, with some cultures having been better at it than others. Any people keen on writing (Egyptians, Phoenicians, Chinese, Greeks, Mayans, ancient Indians and many others) all had types of formal education. Schooling as we might recognize it today, though, began in early medieval Europe with the monasteries – and also in early Islamic Madrasah which taught medicine, philosophy and science.

By the 19th century, when the better off had been schooled for centuries, day schools became more popular. Principally geared toward those whose parents could not afford to pay for their children to board, day schools were those recognizable places that children attend for instruction during the day but would leave to go home at night.

During the 19th century they became infamous for being utterly brutal places. Corporal punishment in schools is now illegal in the majority of countries throughout the world. Most prohibitive legislation has only entered the statute books in the last couple of decades (though there were a few early adopters, such as Luxembourg which banned it in 1845, and Finland in 1914). In the early 19th century, a law was passed in Britain making it illegal for schoolmasters to beat their charges to death.

In Victorian schoolrooms, classes might include up to 300 pupils. The rooms had high windows, so the children would have the minimum of distraction (and sunlight). Discipline was a serious business for teacher and pupil alike. Not acquiring information or skills at the same rate as other children was seen as obstinacy or rebellion and so would incur penalties. This meant that having a learning disability such as dyslexia (another environmentally-expressed disorder) was punishable.

For most of the century the arrangements were all rather informal. This was also the case in other countries. India, for example, had a rather robust educational system in the early 19th century. Like others, it had its roots in religious education. Early gurukuls (traditional schools where children were taught under the guidance of a guru) were among the first public education institutions, relying as they did on public donations. With the arrival of colonial forms of education the number of formal and informal institutions dropped, as did student numbers. (Colonizers have little investment in educating the colonized. Actual education was only really available to the richest tranche of Indian society, much as in Britain.)

Then Dame Schools popped up all over Britain. These were usually run by a single woman, sometimes illiterate, because the point was not really the children's education; only their minding mattered. As the century wore on,

"February – Cutting Weather – Squally" by engraver and illustrator George Cruikshank, from *The Comic Almanack for 1839* (London).

reformers became increasingly concerned, not only about the state of schools, but about the nation's children. The Education Act of 1870 introduced a law stating that every town and village had to provide a school; families paid only a few pence per week per child. A decade later it became law that all children aged between five and ten years old had to be provided with an elementary education.

Similar laws appeared in other countries. In the US, Massachusetts was the first state to make basic education compulsory as early as 1852, with Mississippi being the last in 1918. The dates on which other countries inaugurated compulsory education vary wildly: Japan 1868; Prussia 1763; Greece 1834; Italy 1877; Russia 2007; the Netherlands 1900; Denmark 1814; and India 2009. India may seem to have been rather late to the game, but they now have the largest compulsory educational system in the world. There are still nearly 30 countries worldwide that have no constitutional guarantee of free primary education.

What is so interesting about Victorian schools is their rather unusual methods. They were renowned for providing a highly regimented environment, with gravity-defying levels of boredom in the curriculum: reading (the Bible or a primer), writing and numeracy – and of course a whipping for those a little slower to acquire knowledge than their counterparts.[5]

In Victorian factories, workers went to a place where they were separated from their families, where they were mastered (we would call it managed) by the people in charge, were subject to oversight and planning, where everything was standardized and supposed to be done with the greatest efficiency, and where the outcomes were supposed to contribute to society in some way. Efficiency and uniformity were what mattered most.

There is some coincidence between these things and the rise of schools in the same towns and same period as these factories. Somewhere along the way we turned education into a factory. The instilling of factory rules and discipline into the young was a governance strategy; those born into it would learn to embody their mechanized lifestyle more and more unquestioningly.

This conflation between the factory and the school is not dissimilar to what we see happening in today's education system, where business practices are crammed into schools and universities, but, like the giant foot of an ugly sister being stuffed into a fine glass slipper, no amount of pushing and twisting and straining will make them fit one another.

Charles Dickens knew all this, presaging some of the disasters that result when children are treated as learning machines. In *Hard Times*, he tried to write a novel about factory labour, but without realizing, wrote one about the newly-emerging educational system. The book's famous opening sentence is: "Now, what I want is Facts. Teach these boys and girls nothing but Facts. Facts alone are wanted in life. Plant nothing else, and root out everything else."^

In Victorian schools there were also lots and lots and lots of chairs. Chairs were, and are, an essential part of the disciplinary structure in schools. For most of the time schools have been around, children have been caned for not sitting in them; now they can be excluded and sent to a special school with other similarly disruptive children who refuse to sit down. Discipline is as important in a factory as it is in a school and Dickens completely understood this. Indeed, *Hard Times* opens with a forefinger pointed at seated children, where they are being berated for their predilection toward "fanciful" things. In schools then, and in schools now, pupils must be taught discipline.

Our bodies are obviously different when we are young. The explosive

5 I went to school a century after the Education Act, in the 1970s, and there were still dunce's hats and canes, with left-handers being forced to write with their right hands. I remember being hit in the face for forgetting how to subtract; I think I was six.

A typical Victorian schoolroom. (Note the absence of windows to diminish distraction.)

energy we have, when instead of walking we run – not as adults do, but in an all-out sprint – fades at no single, identifiable point during adolescence. It is as if schools were invented not to educate us, but to teach us to sit still.

By the age of 16 the Anthropocene body has not only been taught to sit still for extended periods, but having done so for so long, the child who went into the education factory limber, supple, energetic and agile comes out the other end with a body possessing a more limited range of motion in many of its joints and limbs. The movement has been lost because of the severe restriction of movement involved in sitting down for such long periods, and has a good dose of obedience inculcated into it, too.

Where do these threads lead us? How is the history of 19th-century education being played out in the Anthropocene body?

The factories are mostly gone now. They are out-of-town carcasses for a way of living long in decline. But the factory model of education is still here. Just as Victorian factory workers were taught to be docile employees, the baton their educational system handed on to us is that we in the 21st century are teaching our children to be sedentary. Our schooling is imparting to them that sitting for long periods is normal and something to be prepared for in adult life.

There has been a surprising absence of inquiry into what exactly our schooling does to our children's bodies.

Research shows that our kids sit down at school for five to six hours a day. Being made to sit still for long periods sounds more like something we would expect to happen at an NSA black site, but when our kids show a completely natural response to this factory-mode of learning and education, we tell them they're unruly, insubordinate and in some cases, ill. This is just one example.

Attention deficit hyperactivity disorder (ADHD) is a neurologically recognized condition which has met some suspicion and a great deal of misunderstanding in recent years. The main symptoms of ADHD are: impulsivity, hyperactivity, lack of focus and occasional aggression. For a majority who share the condition, it is not just about being hyperactive, disruptive and having an inability to concentrate, but more a lack of control over what is concentrated upon. The fact that the environment plays a role in its expression is emphasized by the fact that it is peculiarly sensitive to the place of its expression.[6] Indeed, one of the diagnostic criteria for the disorder is that symptoms must be expressed in more than one place.

In 2004 in the UK, just over 350,000 prescriptions were issued for methylphenidate hydrochloride (Ritalin; the most popular treatment for ADHD). By 2015, this number had jumped to 922,000. A year's use of Ritalin has been associated with stunted growth (by 2cm/¾in), with some evidence suggesting that it can also lead to self-harming.

A disproportionate number of American children (6.1 percent) are treated for ADHD by being placed on drugs.[7] Males are three times more likely to be diagnosed, with the hot zone for diagnosis being children in the Midwest (the states of Louisiana and Kentucky both treat more than 10 percent of their children). Genetics alone cannot account for such disparate numbers. These children's bodies are rejecting sedentariness and almost our first response is discipline, expulsion or a psychoactive pill.

The foundations for this disorder were laid down in the 19th century

6 Environmental psychologists have run trials and parents report that certain kinds of activity exacerbated symptoms (TV and video games, for example).^ Unsurprisingly, energetic play in a green (natural) outdoor environment (as opposed to an indoor or urban one) was most effective at relieving symptoms.

when children were brought into schools and required to sit still. In previous centuries, the genes associated with novelty-seeking and risk-taking behaviour would have been expressed in people such as Marco Polo, Magellan, Columbus, Shackleton and countless others.[8]

In evolutionary terms, any number of explanations would favour the selection of risk-taking behaviour: for example, it might be sexually attractive, or at the level of the tribe it might have been positively desirable to have a small number of individuals willing to take risks and explore new territories. This sense of adventure was not pathologized; it was celebrated because it was necessary.

Today in the Anthropocene, with the roads all constructed and phones to map our way, the inheritors of these genes in the Midwest are often drugged out of their impulsivity and curiosity, taught to sit still and never allowed to get within arm's length of a boat for fear they might take the rudder and steer it into the pier.

---

7 In 2010, a research paper presented evidence showing that those children who were youngest in their class were also the most likely to be diagnosed as having ADHD, "suggesting that many diagnoses may be driven by teachers' perceptions of poor behaviour".^ The age at which ADHD most commonly shows is seven (when being almost a whole year younger than the eldest in your class is a comparatively large margin).

8 Some of the key genes associated with ADHD are CDH13, TAAR1, MAOA, COMT and D4DR.

CHAPTER 6

DEVELOPING (BAD) WORK HABITS

My body is damaged from music in two ways. I have a red irritation in my stomach. It's psychosomatic, caused by all the anger and the screaming. I have scoliosis, where the curvature of your spine is bent, and the weight of my guitar has made it worse. I'm always in pain, and that adds to the anger in our music.

Kurt Cobain

All sedentary workers suffer from lumbago.

Bernardino Ramazzini

For many in the 19th century, work was perhaps even more precarious than it is today. If you were not fit to work, your family stepped up to help if they could and if they couldn't you were out on the street, either looking for cheaper lodgings, or at the very worst taking up begging in order to avoid debtors' prison or the workhouse.

Tailoring was one of the larger unregulated industries, and for those at the bottom of the laundry pile (many thousands throughout the country) jobs mainly consisted of repurposing or recycling existing garments. It was hard work. Life at the other end of the needle often left them in dire need, enjoying none of the required minimum conditions such as meal breaks or sanitation. Their pay was reliant on a market over which they had no control. In the wider market, with many all too willing to work, wages were a race to the bottom. which meant ever longer hours and ever poorer conditions. In 1863 Mary Ann Walkley, indirectly employed by the royal court, died of overwork; the

20-year-old seamstress had squinted at her needle, hunched and pinched, for 26½ hours without a break before dying of exhaustion.^

Tailors in particular were renowned for two predicaments; going blind and sustaining back injuries from their work. Almost every part of their job involved being hunched over, whether leaning over a work table while measuring and cutting material, or sitting cross-legged sewing. After one or two decades in the rag trade, they were often left disabled with a crooked back.

Friedrich Engels in *The Condition of the Working Class* (*see* page 137) noted that in the factory cities and towns, "Malformations of the spine are very frequent among mill-hands; some of them consequent upon mere overwork, others the effects of long work upon constitutions feeble, or weakened by bad food. Deformities seem even more frequent than the diseases."

At the more physical end of the scale manual labourers, stevedores working on the docks and men on building sites were all exploiting their natural abilities at full capacity; they were often injured when they went beyond it. Stevedores were lucky to be given the work after queuing for hours at the dock gates; they lugged sacks on and off ships, and onto transports, and suffered injury rates as high as 50 percent. As the principal breadwinners (which, given their wages, is a term that takes on particular resonance), if these men were injured, it spelled disaster. It was not until the end of the 19th century that their circumstances improved after massive and sustained strike action.

Today, according to *The Lancet*'s "Global Burden of Disease Study of 2015", back pain is the biggest single cause of global disability.^ While statistics such as average global life-expectancy had increased by ten years since 1980, the rise of non-communicable diseases and morbidity suggested that these were not "good" years (years without chronic disease, pain and immobility). Instead, they were dominated by limb pain, complications from osteoarthritis, type 2 diabetes, dementia, stroke and heart disease.

More than four out of five of us who live in the West will experience back pain at some point in our lifetime.

But the numbers, as always, don't tell the whole truth. Mean age and rates of back pain are connected. The reasons our populations are getting older are manifold, but one huge factor is the drop in deaths from communicable diseases. So as our populations grow older they become increasingly likely to suffer with back pain when osteoarthritis strikes and the ability of our lumbar discs to deal with impact, stress and fatigue reduces. For the same reason, there is a reported spike in cancers, Alzheimer's and other dementias, too: there is an increasing likelihood of their being expressed as we age.

Countries where people suffer most from back pain include Japan, the US, Germany and Poland, whereas the numbers suggest that those who live in Angola, Kenya and Ghana are the least likely to suffer. If we cross-checked for rates of cancer or heart disease in these countries, the answers would be similar. Those African countries start to look super healthy. The reason for such good numbers, though, is that life expectancy in those countries is shockingly low. Morbidity has less opportunity in an environment where people are not expected to live into their forties. An additionally strong cultural component is the fact that with no one to listen, little access to healthcare (because of both its accessibility and its cost), the data is unlikely to be forthcoming.

Given human back problems during the Victorian period, and even worse ones today, might it just be the case that we are badly designed: that the spine is an ideal structure for quadrupedal locomotion, but not bipedal? Are we genetically predetermined to suffer back pain?

Evolution would not have great influence on the subject because as a variation, the likelihood of experiencing back pain in older age would have little impact on procreation or sexual selection, so the pathology is unlikely to be deselected. Are our backs the greatest design gaffe since the Sinclair C5?

Comparing ourselves, particularly our spines, with those of other primates reveals what our bodies were best adapted for, and how we came to be who we are. One of our closest cousins, the gorilla, has a back that is straight as a table top. The work it is required to do is substantially different from that of humans. Instead of forming a supportive bridge between the arms and legs, the backs of bipedal humans have to do additional work. While the two structures share overwhelming similarities, their differences are highly informative.

The ribcage of a gorilla is much more barrel-shaped than humans. Ours needs to be flatter to maintain a good centre of gravity. Our spines also need to be effective shock absorbers (hence the S-shape, which also helps the centre of gravity). The mechanism our spines connect to is much larger comparatively. Our pelvis, unlike those of other primates, has to support our body weight, and it also has to play a role in supporting the organs in the abdomen, as well as stopping the body from tilting laterally (to the side) during the gait cycle when walking or running. Shock absorption is particularly important when running.

A remarkable 2017 study from Deakin University in Australia found that "Running exercise strengthens the intervertebral disc."The title is only slightly misleading, as the study found that walking did, too. Previously, there had been no evidence to suggest that our intervertebral discs responded positively to the wear and tear of running. But this study of women and men who had run a minimum of 19km (12 miles) a week for more than five years, found that their disc composition was stronger (with more proteoglycan), their discs were better hydrated (which helped them absorb pressure and impact), and they were larger in volume (better at dispersing impact). These advantages were particularly noticeable in the lower lumbar region, the area most commonly associated with back pain. The even better news was that this was also the case for the set of "brisk walkers" tested.^

> To better understand what kinds of activities generate these acceleration magnitudes, we collected additional accelerometry data under different conditions. Walking or slow running at 2 m/s fell inside this range [that stimulates disc growth] with slower walking falling below this range. Fast running and high-impact jumping activities were above this range.

For the purposes of this exploratory study, the brisk walking was seriously fast, at 13 minutes 24 seconds per mile. The study represents a great breakthrough in knowledge about our spines, as it is ongoing evidence that it is possible to change the health of our intervertebral discs for the better, as well as for the worse.

A number of anthropologists have blamed backache on the fact that we are bipedal, claiming that it is an inevitable outcome of certain kinds of load being placed on such a structure. And while there is clearly a strong argument for this, there is not a single scientific paper that shows that walking is bad for us. Not only is it good for the back, but it is associated with a cascade of related physiological, biological, psychological – even environmental – rewards. It's a miracle cure for hard and soft tissues alike.

The spine is made up of several parts. The inward curve of the lower back is called lordosis; the outer curve kyphosis.

Name	Number of bones	Where on the spine
Cervical	7 bones	neck
Thoracic	12 bones	ribs
Lumbar	5 bones	waist
Sacrum	5 bones fused together	pelvis
Coccyx	4 bones fused together	tail

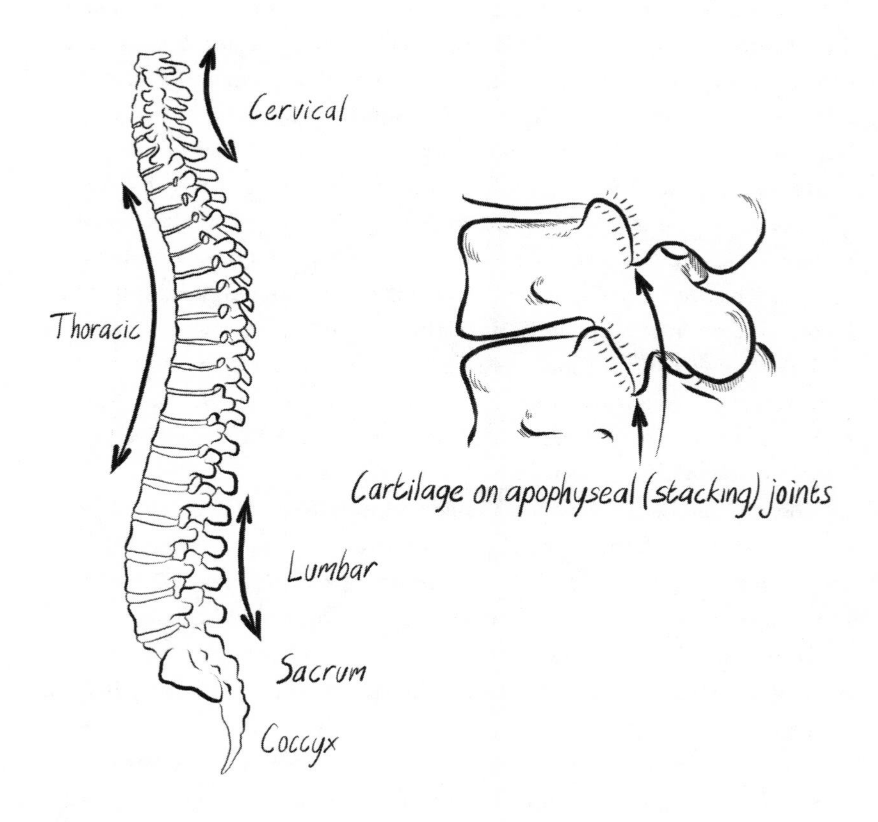

A spine with typical lordosis in the lumbar region and kyphosis in the thoracic.

Half a billion years is a long period of evolutionary experimentation, and there are still substantial differences between our spines and those of other primates.

Depending on their posture, gorillas have either a straight spine or a C-shaped one. The shape of their lumbar spine is very old, and is found in a variety of other species, whereas ours, curving in the opposite direction, is new.

The other key differences between us are that gorillas have a tall and narrow pelvis; the human pelvis is bowl-shaped. Our arms are much shorter, too, with our legs articulated directly below the body (a gorilla's move laterally away from the body when they walk upright). When a gorilla is bipedal, their centre of gravity is somewhere in the soft tissue of their stomachs. In humans, it is at the centre of the pelvis. This means that our spines enter our skulls over our centre of gravity; our knees align over it, too.

Being bipedal, our spines have to cope with considerably more compression than most quadrupeds; this is reflected in the relative size of our intervertebral discs and the width of our vertebrae. The size of the discs is part of the problem: the larger they are, the more difficult it is for them to be repaired because more of the disc is inaccessible to metabolites.

This shows us that our spines are well-adapted for bipedal movement, and indeed the research shows that they seem to rather like it. So what's going on? It's becoming increasingly clear that our spines have not evolved to do certain kinds of work.

There would have been a full alphabet of work to be done in the millions of years before the Agricultural Revolution 10,000 years ago. But once cereal crops became a principal food source, things began to change. The work of provision began to take on rhythms that had not been seen before.

By the time of the Industrial Revolution, the sheer amount of work being done began to accelerate certain kinds of pathology (especially back problems).

During the mid- to late-Victorian period, those at the lower end of the scale would spend the majority of their waking hours working, and – almost uniquely in the history of our species – this work was repetitive on an unimaginable scale. This period also became a stepping stone toward introducing increasing refinement of the ranges of movement that the working body was likely to experience.

We are from the grasslands; that is where human bodies evolved and developed, as distinct from those of other primates. Our anatomy is built to be bipedal, but obligate bipedalism is not to be confused with the idea that we must always be on our feet for longer and longer periods of time. Looking at the legislation and statistics from the 19th century, it seems that the Victorians believed otherwise.

BACK-BREAKING HOURS OF WORK

It took time for human bodies to collapse in the ways that many did during the 19th century. Just as we do today, workers had to invest heavily in sedentary behaviour for muscles to atrophy and bones to thin. Long spans of time were needed for this, and during the 18th and 19th centuries, the working days that were pathological to workers' bodies only seemed to be getting longer.

It is impossible to write meaningfully about the length of a working day in early cultures. Whether in the early cities that were built by slave labour, or where the peasantry worked the land, there was never a need to record how many hours a day people worked. During the new Victorian period of hyper documentation, the length of a working day had never been as forensically inspected over a sustained period as it was in the legislation of the 19th century. Tens of bills were passed, sometimes several a year as the legislation coughed and sputtered its way onto the statute books, and it threw these hunter-gatherer numbers (of about 30 hours a week) into sharp relief.

The legislation table (*see* page 158) gives only a sense of what was going on in workplaces. It's also important to remember why such legislation existed. It is easy to assume that once these acts were passed they marked the end of a particular brand of exploitation. We have laws to protect equal pay in the workplace, not because the problem no longer exists, but because unequal pay is omnipresent. These acts did not signal the end of such practices, but their prevalence. So common were they in fact that the practices required the gathering of Parliament for debate so that legislation might be penned.

These new laws were related to other legislation (such as the Education Acts), which was reflected in the laws passed to protect younger workers toward the end of the century. There was also the Second Great Reform Act of 1867 which widened suffrage to include all men over the age of 18. After a number of fever outbreaks, the Sanitary Act of 1866 also prescribed minimum standards of cleanliness in workplaces to a chorus of contemporary objections. Why was Parliament wringing its hands over the conditions of young children earning a wage, when under-tens in other professions, or the unemployed (unemployed!) were much worse off?

While it might appear that the excesses of the factory owners of the 19th century were tempered by this legislation, much of it (especially in the earlier part of the century) was unenforceable because it was left to a local magistrate to administer. He could only inspect a mill if he had evidence that the law was being broken, and even then, was required to recruit two witnesses for the inspection. When complaints were brought, particularly earlier in the century, mill owners who were magistrates were permitted to adjudicate on their own cases.

A SELECTION OF VICTORIAN LABOUR LAWS

Name	**Purpose**
The Combination Act 1799 (prohibition)	Prevented workers combining and organizing; banned trade unions; repealed 1824
Health and Morals of Apprentices Act 1802 (hygiene)	Health and welfare of children in factories
Cotton Mills and Factories Act 1819 (working hours)	No children under 9 to work; those aged 9–16 limited to 12 hours a day (cotton industry only)
Master and Servant Act 1823 (prohibition)	Became a crime to quit a job without notice (not fully repealed until 1889)
Cotton Mills Regulation Act 1825 (working hours)	Made inspections easier (mill owners complained)
The Combination Act 1825 (prohibition)	Permitted workers to organize, but striking was illegal
Act to Amend The Combination Act 1829 (working hours)	Eased the path to serve mill owners with legal documentation
Hobhouse's Act 1831 (working hours)	Night work in mills forbidden for anyone under 21
Labour of Children in Factories Act 1833 (working hours)	Child labour limited to a 10-hour day in factories (mining *et al* and other work excluded); many mill owners complained
Mines and Collieries Act 1842 (working conditions)	Prevented women and children working in mines

Factory Act 1844 (working hours)	Extended 12-hour and night-working ban to women
Factory Act 1847 (working hours)	10-hour working day for women and children
Factory Act 1850 (working hours)	Women and children to work 6am–6pm, all work to end at 2pm on Saturday
Factory Act 1856 (working conditions)	Relaxed fencing requirements for moving machinery established in 1844 act
Factories Act Extension Act 1867 (working conditions)	Extended to non-textile industries and workplaces employing 50 or more
Factory and Workshop Act 1878 (working conditions)	No child under 10 to work, 10- to 14-year-olds to work ½ days, women up to 56 hours a week
Factory Act 1891 (working conditions)	Prevented women working within four weeks of childbirth; minimum age of workers raised to 11
Factory and Workshop Act 1901 (working conditions)	Minimum age raised to 12, meal breaks and fire escapes
Factories Act 1961 (working conditions)	Consolidated previous acts of 1937 and 1959; laid down health and safety regulations in workplaces
Health and Safety at Work Act 1974 (working conditions)	Updated 1961 Act

The 1844 Factory Act put the brake on extending working hours for the first time, and also required dangerous machinery near women and children to be fenced off. Women and children were not permitted to clean moving machinery. Accidental deaths had to be reported and investigated. Mill owners who were also magistrates were prevented from hearing cases relating to the Factory Act if they, or any member of their family, were an interested party.

The two principal voices objecting to industrialization in the period were the sage and social critic Thomas Carlyle, and later the art critic John Ruskin. Both men used gallons of ink writing in defence of the working poor, but also in thinking through the damage being done to their crippled bodies, to those who employed them and to the environment.

The idea that there was a "Condition of England Question" was Carlyle's, first used in his book *Chartism* in 1839. In his other works, *Signs of the Times* and *Latter-Day Pamphlets*, he critiqued the exploitation of the working classes, and in *Past and Present* he called out these new "captains of industry" as being more like a "gang of industrial buccaneers and pirates."^

John Ruskin, though much better known as an art critic, often concerned himself with the conditions of the workers, making the art and architecture of his concern. In *The Stones of Venice*^ he espoused the love of gothic architecture because he could see that in its construction the minds of the workers had been exercised – Ruskin was indirectly praising the value and inefficiency of the manual work that he could see in the finished product. He liked the imperfections of this kind of architecture because he found humanity in it. "Imperfection is in some sort essential to all that we know in life."

Later, William Morris, an artist and designer who turned increasingly to politics, characterized the new class system as, "a class which does not even pretend to work, a class which pretends to work which produces nothing, and the class which works, but is compelled by the other two classes to do work which is often unproductive."^

While the majority of adults and children who worked did so in factories for much of the century, those in other less organized trades were left to fend for themselves. Those in domestic service probably worked the longest hours of any cohort in employment, and there were no government inspectors to check on their hours or conditions. There were also many thousands of unnamed workers who banded together in the slum districts of the rapidly-growing cities to ensure that jobs were done on time. Often whole families mucked in together to complete an order by a given deadline so that they might share the few pence that was paid for the work.

William Dodd in *The Labouring Classes of England* showed how work consumed everything and everyone in its path.^ In one chapter, "The Lace Makers of Nottingham", he explains that the boys who worked the bobbins were beaten with canes when they looked as though they might fall asleep from exhaustion; the girls had bodies requisite to the project, too, because their "delicate little fingers" were necessary to shape and work the lace. The mothers who dosed their babies with laudanum, leaving them "pale, tremulous and emaciated", did so that they could work in peace. All did what was needed to get the job done – giving more than any of us might.

The working day had lengthened. Despite efforts by the government to control abusive practices, there was no law to control the kind of work that went on by the hearth, or in any number of other industries.

WORKONOMICS & ERGONOMICS

Occupational injury has existed for the same length of time as occupations. *Homo habilis* and *Homo erectus* would have injured themselves shearing flints in their tool manufacture as easily as the clerks in counting houses would have suffered from a bad back after being seated at their workstations for 14 hours a day. Workload minimization really started on the savannah with tool use, but it came into its own with inventions such as the wheel, the lever and the pulley, which permitted the construction of substantial buildings and monuments. There would have been no cities or Stonehenge without these early technologies – and the erection of those monuments was only achieved with the kind of strength that modern humans can only imagine.

There are probably few workers in the history of our species who have not made some effort to optimize their work both in terms of making it more efficient and also in making it easier. The first documented evidence of an interest in ergonomics comes after the fall of Sumeria and Babylon by the Greeks.

Hippocrates (460–370BCE) was the Greek philosopher and thinker whose ideas became the cornerstone of medical thought (it is for him that the "do no harm" Hippocratic Oath was named). Hippocrates was the earliest writer to work explicitly on ergonomics. In *On Hospitals*,^ he focused on the ideal working environment of surgeons and the placement of their equipment:

> As regards the tools, we will state how and when they should be used; they must be positioned in such a way as to not obstruct the surgeon, and also be within easy reach when required. They must be close to the surgeon's operating hand. If an assistant passes on the tools, he must be prepared to pass them as soon as they are asked for.

This is only a snippet of the pages of guidance that Hippocrates gave for the organization and optimization of workspaces in hospitals. There are a great many other guidelines for working practices, too. But Hippocrates was not alone in thinking through how the human body might perform its work best. Throughout the Greek's architectural designs there is evidence of careful ergonomic planning and forethought.

Although the philosophizing over ergonomics does not reappear for another couple of millennia until the mid-19th century, its significance is huge and an

From E W Duffin, *On Deformities of the Spine*, 1848: an example of simple habit correction.

early acknowledgment of an idea that now dominates our engagement with technology and all kinds of work, from the most sedentary to the most active.

For all the thousands of figures on Greek pots, vases and friezes, represented in all manner of poses, there are none of people clutching their back wincing in pain, except for the rare few who have an enormous spear jutting from it. The emergence of back pain was well on its way and would later be accompanied by much hand-wringing from employers – especially those worried that making workers sit obediently in chairs all day may not be great for their long-term health.

While I was in Sardinia with Gianni Pes (*see* page 113), I saw advertised on TV a sort of support garment. It looked a little like a gun holster but it was a posture correction corset. Priced at about 30 Euros ($35), the idea was that people with back pain would buy this in the hope that it would solve their postural difficulties and cure their pain. But, just as with supportive shoes, using such a garment is likely to start a downward spiral. Wearing it is supposed to create a "muscle memory" in the bodies of wearers to enable them to correct their posture, but wearing the corset would weaken precisely those muscle groups that should be recruited to fulfil the task.

Good posture should not be an effort; it should result from a biomechanical muscular system that is evenly in balance. Surely it would be better to strengthen failing structures so that they can support themselves, requiring no further conscious effort at all.

How did we get from the seemingly pain-free – or at least uncomplaining – Greeks, to strapping on support corsets because we can no longer stand up straight by ourselves any more?

If we deploy our literary periscope once again it sees only an empty landscape; there is almost nothing in literary history that concerns itself with bad backs. Just as with the Greek vases, the bad back does not really appear – there are none in Shakespeare, but after that we begin to hear their groans with some frequency. Something changed, then, in the early 19th century which started this plague of disability.

During the 19th century there was a widely-held belief that habit could leave a marked impact on the body. Today, these ideas still persist. We are told to sit up straight and maintain good posture; that problems with our back can be solved by an effort of postural will. Likewise the contraptions that can be bought to help solve these problems appear in the Victorian period.

One of the earliest exercises in popularizing sometimes erroneous healthcare beliefs was Edward W Duffin's *On Deformities of the Spine*, which first appeared in 1848.^The book is a broad and intelligent study of the spine which draws together much contemporary debate. It also contains those two necessaries for any health book – advice that puts the onus for pathology on the individual, and a machine to help cure it.

In the first instance, there is an illustration of two girls working at their desks, the first of whom is in the process of damaging her spine because of "a bad position habitually assumed whilst engaged in writing". The second shows how the use of something as simple as a wedge can straighten her up. "A momentary glance at the woodcuts, will suffice to illustrate the bad consequences that might result to a girl somewhat crooked, whose mode of sitting, whilst engaged at her ordinary studies, is not carefully tended to: and the facility with which these might be counteracted is equally exemplified."

For those who struggle with similar complaints, Duffin recommends lateral exercises as "a mode of using the spinal muscles generally too much overlooked... For such purpose, I have been in the habit of recommending a frame somewhat like the one which a rocking-horse stands, but rather more acute in the curve, and having the ends turned inwards."

What I like about both these examples is their familiarity and that they remind me of the advice we are now overwhelmed by in the media, in the

From E W Duffin, *On Deformities of the Spine*, 1848: an early "home exercise" machine to work the back muscles of those who did little the rest of the day.

gym and when we talk to one another about our bodies. We believe it is our habits and lack of ergonomic methods that have caused our problems, and look less toward our modes of long, sedentary work as being among the most influential practices of all.

The next time ergonomics really comes into focus is when Britain is beginning to legislate against the capitalist indulgences of the Industrial Revolution. The table of legislation (*see* page 158) related to working practices provides only a sense of the range of legislation passed. What these laws demonstrate is a persistent concern with the ways in which the ecology of labour, the range of working practices, had changed for the worse in the period.

In the late 1850s, the first "study" of ergonomics was published by a Polish physician, Wojciech Jastrzebowski, *An Outline of Ergonomics or Science of Work based upon truths drawn from the science of nature*.^ He explained what he meant by ergonomics:

> By ergonomics we mean the Science of Work. Useful work is divided into physical, aesthetic, rational and moral work. Examples of these kinds of work are as follows: the breaking of stones, the playing with stones, the investigation of natural properties or the removal of stones from the roads so that they not give rise to untidiness or suffering for people.

His desire was that humans, with the right help "could work at the least expenditure of toil". The book was not an immediate bestseller, but it is of its moment. It recognizes the damage wrought by repetitive work. Whereas John Ruskin and Thomas Carlyle worried that the worker was being turned into a machine, Jastrzebowski's sights were set a little lower: helping the man to work the machine.

Ergonomics is the science of fitting together several things:

1. Workplace requirements: such as typing at a certain speed, lifting or transporting materials, using telephony or computer equipment.

2. Working environment: this might involve risks (moving machinery, air pollutants, darkness, heat or cold) or require the worker to assume certain postures for extended periods of time.

3. The capabilities and needs of the human body.

4. Productivity.

The first three are far more complicated than they might first appear, with the third surely taking the prize for all-out complexity. The watchwords of ergonomics are "ease and efficiency". The science is about wellbeing, but it is also about optimizing productivity. It's as much about economics as it is physiology. If the employer gets the first parts right, the pay-off comes in the fourth. While streamlining systems and making workplaces more ergonomic might only have a small impact on output; once this is multiplied across an entire workforce, the differences can be substantial.

Businesses could pay more attention to ergonomics. Several studies have shown that following an ergonomic intervention in which new and appropriate office furniture was provided, productivity was measured as increasing at rates of between 12 and 18 percent. In schools, the anthropometric needs of students were assessed and new desks and workspaces were provided. Of these studies 64 percent presented positive results with proven effects on learning and information uptake; 24 percent were negative, with the remaining 12 percent showing no impact. High furniture, sit-stand desks, tilting tables and seats were among the most positive interventions.^

The greatest ergonomic risk factors are awkward postures, extreme temperatures, poor lighting, repetition, vibration, static postures and heavy lifting. The 19th-century factory worker endured all of these. But these also translate into today's more sedentary work for many of us, with the additional risk of contact stress, when sensitive tissue (such as the underarm skin above the wrist) rests on the edge of a desk while typing.

The kinds of manual labour that factory hands were subjected to for 16 hours a day is fundamentally undramatic. The Greeks on vases and pots who are portrayed with spears protruding from their sides with gaping wounds are inherently dramatic; chronic pain is not.

The catastrophic realization that one's life has been wasted in useless toil for someone who has exploited you, destroyed your body, taken their profit and left you to starve, is the single most dramatic moment of a working life. It is for this reason that we need two things: one to do with form, the other, genre; things that the industrial novel simply cannot provide: vision and comedy. We need film to understand the tragi-comic work of the industrial machinic body.

Charlie Chaplin's *Modern Times* (1936) is a comedy classic. One of the reasons it is so good is that it fills us with the horror of recognition at what our workplaces have become since the 19th century and forces us to laugh at this.

In the opening sequence, Chaplin works at a conveyor belt, doing a kind of manual work with an intensity of speed that is both funny and alarming. With a spanner in both hands, he tightens a pair of nuts on a steel plate as they whizz by on a conveyor belt, with more people down the line who have their own tasks to fulfil. This production line is completely disrupted when Chaplin's character tries to do something as simple as scratch his nose or swat a fly from his face. Everywhere, the employees are overseen by a "big brother"-type manager who barks instructions from behind screens – even in the workers' lavatories.

When the machine stops, the movements have worked their way under the skin of Chaplin's character to such an extent that they reverberate around his body and he cannot stop his arms twitching, or his eyes seeing things that his muscles feel need to be tightened: his butch colleague's nipples, the buttons on a woman's dress. His muscles have learned their work and his brain seems to have little authority to stop them finding employment – he has, of course, become a part of the machine. This point is fully made when Chaplin's character falls into one of the giant frames and his body processes along a route made up of cogs and pistons that he seems to have become one with.

In another sequence, the owner of the factory wants to try out a new time-saving device so his workers cannot slack on the production line. The owner is rather taken with the salesman's pitch. He tells him: "Don't stop for lunch; be ahead of your competitor." Instead of taking the usual meal breaks and wasting valuable energy feeding themselves, the factory trials a machine that fits around the worker's body and shoves food into his mouth. Comic, yes, but also portraying the inherent difficulties of trying to turn a human into an engine, or reducing him to something that needs only to be oiled like a cog in a machine.

What the film makes clear is that factory labour, and indeed much of our labour today, is fundamentally about mining and sourcing efficiencies, for the simple reason that if you take the variety out of any mode of work it can be completed more quickly and cheaply. The operator or hand on a production line can become highly specialized at performing only their short task without the need to acquire extra knowledge or skill during the process. Not only are goods produced more quickly, but the labour, being less specialized, is both cheaper and requires less investment on the part of the employer in training costs. The efficiency gain is usually matched by an experiential loss on the part of the worker.

In different ways, as we negotiate the landscape of modern work, Chaplin's character is the one humans have become, as we sweat over productivity, are badgered by our bosses to "produce and produce", to become ever more efficient, and as we use more and more of our free time to get ahead of our competitors. As the sedentary workers sit there, they might reflect on their posture. It is something every sedentary worker is asked to do at least a few times during their career, through emails, workstation assessments and those postural posters. We have all seen the diagrams of the ideal desk set-up, and the perfect posture while sitting at them. The truth is, though, that nobody really knows what perfect posture is.

Asking a back expert to define "good posture" is like asking a podiatrist to define overpronation. There is no ideal the science has fixed upon that is measurable and identifiable. We can't look at Michelangelo's David or the Mona Lisa as ideal examples of good posture because we don't have their bodies. Posture may be as specific and personal to an individual as their fingerprint or their DNA.

What we can say about posture is that it should be comfortable, pain free and unlikely to lead to pathology.

The first two are easy, the latter is a bit more complicated. Pathology is connected most clearly to immobility; good posture, if it exists, is about mobility. This is why there is no image or drawing in which good posture can be delineated, precisely because it is not static. Posture is a series of micro-changes of angle, vector, weight distribution, lean, rotation, contraction and extension. So many variables are involved, in fact, that it is unlikely that there has yet been a point in your lifetime in which you have held exactly the same posture twice. Good posture encourages movement, whereas the working habits developed in the 19th century promoted the opposite.

MORE ABOUT ECONOMIC BODIES – THE WORK ETHIC & POOR EYESIGHT

The Victorian work ethic we have all inherited has a lot to answer for. The idea developed by the factory owners of the early 19th century that everything should be judged by volume, efficiency and profitability is not only a warped way of looking at life, but is one that is changing our senses.

Poor vision was a conundrum to the Victorians. An article about mesmerism in *The Lady's Newspaper* on 18 March 1848 explained that clairvoyance was as much "a mystery as eyesight or spectacles".

Illustrator and caricaturist Robert Seymour produced a watercolour etching sending up the accoutrements that were necessary to the modern gentleman in 1830: a revolving hat that provided snuff, cigars, spectacles, monocle and hearing trumpet as required.

As the pathology of poor eyesight gathered pace throughout the period, it was first noticed in a specific population among the poorest factory workers. Friedrich Engels, in *The Condition of the Working Class*, observed its spread among textile workers, where he could see that the long hours that low wages imposed combined with "a sedentary life and the strain upon the eyes involved in the nature of the employment" to "enfeeble the whole frame" of the worker "and especially the eyes". Their sight became so poor that "work at night is impossible".

Historically, poor eyesight had strong links to old age. The Roman philosopher, Seneca, having spent a life in letters writing and reading, came

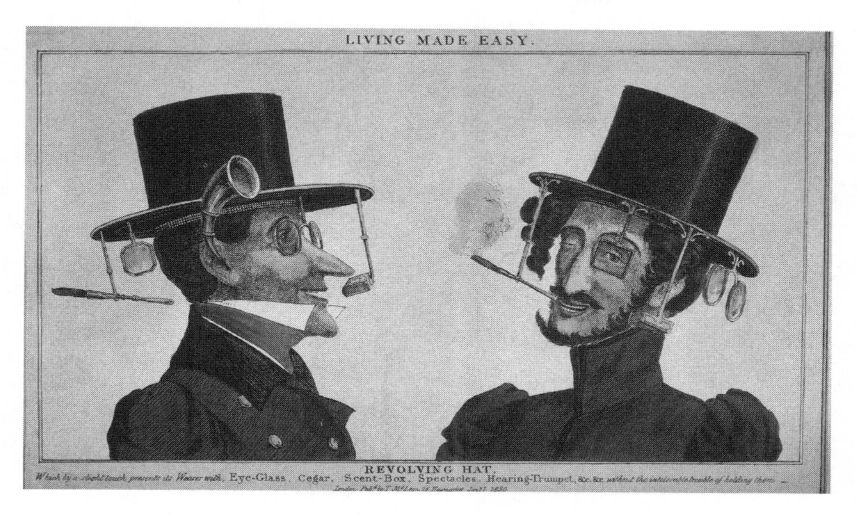

"Living Made Easy: Revolving Hat", 1830, a hand-coloured etching by T McLean poking fun at the accoutrements of modern life and its pathologies.

up with the ingenious idea of filling a glass bowl with water to magnify the letters when his eyesight deteriorated in his later years. That philosopher-monk and early man of science Roger Bacon discussed optics and the wearing of spectacles in his *Opus Majus,* written for Pope Clement IV in 1267.^ Monks and scholars were among the first to adopt eyeglasses, and the reason they required them was most likely because they needed to be able to do close work at a time when their ageing eyes were becoming increasingly long-sighted. This condition, called presbyopia, was what eyeglasses were invented for, being one which gradually worsens with age.

The men pictured by Seymour, though, are not old. And as the 19th century wore on younger bespectacled people become a more common sight. These tended to be well-to-do bookish types such as the dorkish Mary Bennet in *Pride & Prejudice* or Mr Hall, the "accomplished scholar" in Charlotte Brontë's *Shirley.*^ Something new was happening; it seemed that reading and close work was causing a sudden decline in vision.

In the 19th century, spectacles became a badge of honour for the wealthy. They were a facial adornment that spoke of privilege and higher education; an affliction induced by a lifestyle in which all needs were met, and a different kind of hunger was free to emerge and be satiated: the desire for knowledge. At the other end of the scale, for working-class children shunted into schools toward the end of the century, myopia – just like diminutive height – became a marker of poverty. Without the money to spend at ophthalmologists, and with only the most meagre education, this was a population good for one thing: manual work. If we take the hand of this pathology and allow it to lead us momentarily into the present, we will see that the increase in myopia which began during Victorian times is now widespread in high-income nations.

The pathology that gathered pace in the 19th century is now dominating parts of Asia where it is as high as 80 percent among schoolchildren. In South Korea's capital of Seoul, 96.5 percent of 19-year-old men are short-sighted, while in hunter-gathering groups myopia is practically non-existent. In the UK, myopia affects one in three people. In the US, recent figures suggest that it is now up to 40–50 percent of under 16s. But in India, levels are much lower: an estimated 13 percent. In Mexico: 10 percent. Iran: 8 percent.^

Needless to say, myopia is not an evolutionary gift; it is not a favourable condition, not in the least. Its expression in early humans or primates would almost certainly have led to an early death long before the genes had a chance to reproduce. Its recent prevalence must have another cause.

Short-sightedness has genetic components; the condition does run in families; if your parents are short-sighted there is a greater chance that you will be, too. Even though the condition has been connected to hundreds of genes, their sudden appearance in the 19th century, and their more recent expression is happening at too great a pace for the cause to be genetic.

A 1975 study published in the *Canadian Medical Association Journal* looked at an Inuit community in which the condition was only expressed among 1.5 percent of its population. A couple of generations later this had ballooned to just below 50 percent; genes cannot be responsible for this.^

A study by the University of Cardiff in 2015 found that first-born children were 20 percent more likely to develop the condition than their siblings.^ It concluded that the levels of parental investment in the child's education may have had an impact on the expression of their genes. Studies in China have also made a similar connection, finding greater expression in richer provinces than poorer ones, suggesting a comparison emerging between rates of myopia and academic attainment.

Other studies confound this data when children from similar genetic backgrounds are shown to have done similar amounts of near work, with vastly different outcomes for their visual health.^ Even though one of the trial groups did a comparatively greater amount of near work, they reported fewer instances of myopia. The test group with vastly fewer instances of myopia had spent a great deal more time outdoors. Similar studies by the same team dovetailed with these findings, discovering that it was specifically outdoor exercise (not just any kind of sport) that was key. "Higher levels of total time spent outdoors, rather than sport per se, were associated with less myopia and a more hyperopic [long-sighted] mean refraction, after adjusting for near work, parental myopia, and ethnicity."^

Research done as long ago as the 1930s studied 2,364 members of hunter-gatherer tribes in what is now Gabon. Of all the eyes tested, nine would have required between -0.5 and -1.0 correction (exceptionally mild myopia), with five between -3.0 and -9.0. The total rate of myopia, including those who could have got by quite happily without resorting to spectacles, was only 0.4 percent.^

In the UK, the NHS sensibly recommends that because myopia is often triggered in childhood, ensuring "your child regularly spends time playing outside may help to reduce their risk of becoming short-sighted." This is because it is now believed that it is not books, screens or phones that are the most influential causes of myopia, but that it is a by-product of the fact that these things tend to be done indoors[1].

What conditions did the poor and the bookish in the early to mid-19th century share? The common assumption was "close work"; odd, given that little factory work could have been characterized as such. In fact it was that neither group had much access to daylight. The roofs over our heads and the time we spend under them are the main cause. The idea that time must be spent productively is the major driver in the 19th century.

Being indoors inhibits dopamine release which is essential for the development of children's eyes and the neurotransmitter is activated by exposure to sunlight. Dopamine is involved in some of the transformations that take place in our bodies when we switch between day and night modes and although the science is still young, it is believed that there needs to be more clarity in this cycle for young people because dopamine is essential for healthy eye development. As understanding of the light-dopamine hypothesis

---

1 Babies are all born long-sighted. As we grow, the eyeball changes shape and, like an auto-focus on a camera, tries to find the optimal point of refraction. Any camera-owner knows what it is like when the lens can't find its focus – the picture taken, which represents the camera's best guess, comes out blurry and indistinct. The developing human eye in a child is a little like this. Without the right kind of powerful light, the eye cannot learn where its ideal focus is, and ends up by making a similar "best guess" which has little chance of being perfect.

develops over the coming years, it seems likely that exposure to sunlight might become one of the most important preventative medicines for children.

In 2016 in the UK, a report was published that concluded that 74 percent of British children suffered lack of access to green spaces and screen addiction, as a result of parental fear, and now spent less time outdoors than prison inmates. The survey questioned 2,000 parents of 5- to 12-year-olds and found that 20 percent of children did not go outside at all on an average day.^

According to another government report published in 2016, more than one in nine children in its test group had not been to a park, beach, forest or any natural environment in a year.^

Myopia is a classic mismatch disease: a genetic predisposition that had little opportunity of expressing itself in other (earlier) environments, and put down roots in the Victorian period, but is taking over our principal sense in the Anthropocene.

Using data from 145 studies involving 2.1 million participants, a 2016 meta-analysis published in *Ophthalmology* concluded that if current trends persist unchecked, what began with a few indoorsy types in the early Victorian period, and developed through the kids shunted into factories to help support their families, descendants of those behaviours will lead to more than five billion being short-sighted by the year 2050 – that's half the world's population.^

CHAPTER 7

EARLY AIR POLLUTION – OR THE BIG CHOKE

"Ah! my poor dear child, the truth is, that in London it is always a sickly season. Nobody is healthy in London, nobody can be. It is a dreadful thing to have you forced to live there! so far off! and the air so bad!"

"No, indeed – we are not at all in a bad air. Our part of London is very superior to most others! You must not confound us with London in general, my dear sir. The neighbourhood of Brunswick Square is very different from almost all the rest. We are so very airy! I should be unwilling, I own, to live in any other part of the town; there is hardly any other that I could be satisfied to have my children in: but we are so remarkably airy!"

Emma, Jane Austen (1816)

The changes the 19th century introduced to the human body were profound. All around the Victorians was a fluctuating environment, and with it came new ways of work, social interaction, travel, movement, living, learning, earning, developing, playing, dissenting, understanding, thinking and even worship. The environment began to change substantially, from how places looked and felt to the very air that was inhaled by a populace. Retaining an ability to breathe the air should never be counted among the accomplishments of a human.

Although it was the centre of commerce for a growing empire, London was also leading the world in terms of air pollution by the mid-19th century. Economies were growing fast in other countries (such as India) during the period, but the uptake of industrial activity was much slower for a whole range of geo-economic reasons. London was packed full of people, was filthy and stank. Transport, in particular, threw tonnes of dust and detritus into the air and reduced some streets to quagmires in rain. Mixed with all the coal debris in the atmosphere, dust and smuts sometimes floated about, looking and behaving like deathly black snow.

Londoners who had an existing complaint found that the fog triggered attacks and – well – killed them. In the same way as today, the right atmospheric conditions were needed for fog to form, and once in place, it could prove deadly for thousands.

The prophet of 19th-century population growth was Thomas Malthus. His 1798 *Essay on the Principle of Population*^ proved hugely influential. In it he discussed the various cycles of human productivity and how they influence one another. For example, as production methods for growing food improved greater yields resulted, but Malthus believed the effect of abundance would be temporary because population growth would soon catch up with new production methods.

Humans, it seemed, had a propensity for translating prosperity into an ever greater number of mouths to feed. Population, by its nature, would always expand to meet available food resources – just as the height of species (such as *Homo floresiensis*) was stunted because it had practically exhausted its local biocapacity. In the context of the Industrial Revolution, though, things were a little different. Here the lower classes starved or died from complications connected with poorer diet. For Malthus, society was not perfectible, and riches for all were not necessarily the outcome of collective hard work.

There are material limits to the potential for growth and engrained inequalities were most exposed when scarcity struck.

A curious break with the ways energy had been harvested in the past began to emerge as the Industrial Revolution gathered momentum.

While agricultural work had for centuries consumed the energy, loyalty and the time (and lives) of the labourers, when extra capacity was needed, workers had to work longer hours or more workers were drafted in, which meant that extra supplies were needed to fuel their bodies. Workers ate more. More food was produced, which meant more fields had to be ploughed by horses. Horses needed more food, and that food was transformed into glycogen: muscular energy used to create more food, with the by-product of manure, which also fertilized the soil, which grew the food, which fed the horses, which worked the fields, which also fed the population. It is much more complicated than this, but there is an observable cycle of energy return.

In the manufacturing industries, some of this work was off-loaded to machines. Men and horses had to be fed to plough the fields, and then machine energy was added to the equation. Initially, the turn toward fossil fuels in this period appeared to many to release them from the Malthusian problem. Fossil fuels could be utilized to produce an excess of energy for human production. But unlike the horses that plough the field and produce both food and manure (a cycle in which equine energy continuously returns to itself) fossil fuels, once burned, stay burned. Mined coal stays mined. They never return to their previous state. Energy is consumed, and it produces, among other things, smoke that blots out the sun and is poisonous to breathe.

BREATH-TAKING ECONOMIES

Cities are like cells: they don't contain everything they need. Energy has to be drawn into them so it can be used and consumed. Just as with cellular respiration in the body – in which food is consumed and with the help of oxygen broken down in the cell into energy and its by-product CO_2 – cities utilize a transport network to bring fuel into urban centres to be burned and converted into energy and CO_2.

The connection between pollution and the damage done to the environment during the Industrial Revolution seems at first glance blindingly obvious. A surprising fact, though, is that air quality in London has been terrible, not since the 1950s, but on and off for close to a thousand years. While there was a slew of lung-protecting legislation in the 19th century, this wasn't the first time the capital had resorted to such measures in environmental emergencies.

Elsewhere, the blackened lungs of mummies excavated in Egypt tell us that wood smoke in homes caused problems there. In Ancient Rome, the population was beset by the foul air emanating from homes and workplaces. And Roman law courts were considering complaints about smoke pollution 2,000 years ago.

Coal has been mined in the UK since about 3500BCE, but coal mining has a chequered history; it was not taken up widely by any culture before the Industrial Revolution. In Britain the Romans exploited some of the coal fields of the north of England and Scotland, but after they left in 410CE there is little evidence of coal burning for about 800 years, after which coal use suddenly increased in the capital.

At the time of the Norman Conquests in the mid-11th century, England still had extensive woodlands. The Domesday Book and a variety of other sources tell of woodlands located near London and near towns and villages. But in the 200 years that followed there was substantial felling of woodland across Europe and in England in particular.

At the end of the 13th century London was the size of a large town. Shakespeare's Globe Theatre was still in a distant future, but the Anthropocene had already begun. The metropolis was in the process of breaching its full capacity.

As early as 1285, a royal commission was appointed to investigate the use of energy in lime kilns which had previously burned wood but had now switched to substance called sea coal.[1]

The commission found that the fumes produced were much complained about by local inhabitants, and even by those passing through the area. A second commission convened in 1288 with similar results. A few years later in 1298 a coterie of London smiths voluntarily agreed to cease using coal at night because of the unpleasant and noxious fumes it emitted. A few years later, in 1306, Edward III issued a royal proclamation which prohibited craftsmen and artisans from using coal in their furnaces.

Many were not obedient to their king, but despite this not much else is recorded about London's pollution. Except for a couple more references in the following few years, that is the end of the problem in historical records for nigh on 250 years. Which, considering the urgency with which the problem was treated when it arose, is a rather peculiar, if not staggering, silence.

In the preceding years, London had needed a great deal of wood. How much exactly is impossible to calculate because trees vary so much in size – nearly as much as the size of households does – but a medium-sized home would have consumed approximately 15 cords of wood per year (a cord is a bundle about 2.5 square metres or 27 square feet in size) so, two to three per person would have been needed.

Each person would have consumed a couple of medium to large trees, or up to a hundred saplings. Multiply this by the population of London in the

1 No one really knows why it was called that, but as a substance it was a substandard, dirtier and faster-burning energy source than traditional coal. It also contained more harmful elements, such as large amounts of sulphur.

middle of the 14th century (an estimated 40,000) and the annual numbers become astronomical. London required the felling of approximately 120,000 trees annually to keep its chimneys smoking. (It is little wonder that woodland cover in the UK dropped from well over 90 percent around 2,000BCE to 4 percent by the Industrial Revolution).^

On the one hand, woodland was desperately needed to supply London's fuel, while on the other, those woodlands also had to be cleared to make more space for pasture and arable land to feed the Londoners. An alternative fuel source must have seemed a gift from heaven.

In 1661, the famous diarist and Fellow of the Royal Society, John Evelyn, noted the effect of sea coal. So bad was it that "men could hardly discern one another for the Clowd ... that Hellish and dismall Cloud of SEA-COALE," being an "impure and thick Mist, accompanied with a fuliginous and filthy vapour, which renders them obnoxious to a thousand inconveniences, corrupting the Lungs, and disordering the entire habit of their Bodies."^

John Graunt, a fellow of the Royal Society, also noted that before the 1600s death rates in the country and the city were comparable, but in as little as a few decades London had become "more unhealthful" because coal was now universally used. So much so that most people "cannot at all endure the smoak of London, not only for its unpleasantness, but for the suffocations which it causes".^

Why the sudden atmospheric turn? Why such a long gap between the complaints about this irritating new metropolitan cohabitee?

Between the time of the Domesday Book and the mid-14th century, the population of England grew relatively quickly from 1.1 million to nearly 4 million in the earlier half of the 14th century, with London alone doubling in size from 20,000 in 1200CE to 40,000 in 1340CE.^

A sharp-eyed historian (such as William H Te Brake, who first noticed this pattern) would by now have noted the correlation between these dates and a significant depopulation event that arrived in England in the 1340s: the plague.

By 1374, the 4 million people had already been reduced to 2.25 million, and by 1430 the number had shrunk again to 2.1 million. Farms and homesteads were abandoned. Arable fields were left to fend for themselves and returned to woodland in a matter of decades. Less building was also required, not only for the shrunken population, but because of the wide availability of buildings. It is not possible to imagine a more unpleasant environment than London during this period, saturated in death and disease; yet it was a city living in environmental equilibrium. The 250-year dip in the English population corresponds with the time in which complaints about the burning of coal practically disappeared.

But fuel scarcity returned with a vengeance in the mid-17th century. The woodlands were disappearing at a rate never before witnessed. If forests are the lungs of the planet, in Britain they had become tiny gasping husks. In an effort to keep the energy flowing, laws were passed in the 16th century to prevent the export of coal as it was needed so badly for the British economy.

It is alarming to think that the solution to England's first environmental crisis in the 14th century was the sudden death of nearly half its population. But in this period, a scattering of artisan shops were the main offenders, each with a relatively small furnace, narrow chimney and thin whispering plumes of smoke. Fast-forward a couple of centuries and there were great steam-powered factories that produced tonnes of polluting air as quickly as they could spin and weave a cotton shawl.

Where using wood had put a natural constraint on the limits of production, switching to fossil fuels engaged the rocket boosters, cut the brake lines, floored the accelerator and industries squealed with glee as their production crested the peak and began careering downhill at top speed.

Sea coal was still burned in the 19th century. It was used by those on the precarious edges of the supposed economic prosperity of the mid-Victorian period. Most of the coal burned at that time was a little cleaner, but its volume vastly multiplied. So much was mined that it was exported all over the world, and it wasn't until the 1870s that other European economies caught up and competed.

Germany began to vie for industrial supremacy toward the end of the century. The Germans had learned lessons from British industry and were able to spend less on experimenting with new methods and processes. Instead, they invested more in research while expending less defending and maintaining a worldwide empire. This gathering of pace in Germany, as in other countries, was reflected in the quantity of coal they mined. In 1850, for example, a German mine might have produced 8,000–9,000 tonnes annually; by 1900 that figure was closer to 300,000 tonnes.^

France boasted similar figures. In 1820 it was producing in the region of a million tonnes of coal each year; by 1860, that figure had risen to 8.3 million.^

Back in the UK, which had the worst pollution on the planet, a number of "clean air" acts were passed, but there was no halting the domination of fossil fuels as factories sprang up in new cities all over the world. By the dawn of the 20th century, the respiratory disease bronchitis was Britain's biggest killer, not heart disease or cancer, but a lung complaint.^

Fogs and smogs became widespread. The thick, mustard-yellow clouds of the 19th century hung heavily in the air and cast a cloud over the general optimism of the Victorian period. The railway and factory owners complained of the limits placed upon them by governments that gave them, literally, breathtaking freedom to pollute the atmosphere. And the car had not even been invented – though they didn't take long to arrive.

HOMELESS HUMANS

Perhaps one of the most peculiar aspects of air quality in the 19th century is not that pollution made its presence felt after lurking about the capital for nearly a millennium. Instead, it was the fact that seemingly clean air had become an irritant, which was eventually recognized as something more serious.

As early as 1828, writers and investigators such as John Bostock (1773–1846) noted that there was a new affliction that was "mostly confined to the higher ranks of society". Bostock claimed he had scoured the country to find sufferers of this as yet unnamed disease. The paper reported:

> The number of cases which I have either seen or of which I have received a distinct account, amounts to eighteen [...] They all agree in the complaint making its appearance at the same season of the year, in its seat being the membrane lining the nose, the fauces, and the vesicles of the lungs, and, for the most part, in the paroxysms being excited and the symptoms aggravated by the same causes.^

There were probably more than his reported 18 cases in the entire country, but there is no doubt that the disease was rare. One of the first times it was mentioned by name was in Benjamin Disraeli's novel, *Sybil, or the Two Nations*.^ The novel is set in the 1830s when the king, William IV, was in poor health and much gossiped about in high society. The excuse the court gives to explain his absence at social functions is hay fever. What is interesting about the way Disraeli uses disease in the book is how it becomes an acceptable public relations exercise to hide the fact that the king was dying. The novel makes it clear that it was the kind of illness that showed the sufferer in a favourable light. From its earliest appearances the new disease took on a veneer of exclusivity, in the same way as spectacles or lead-whitened skin; ill-properties belonging to the elite.

Charles Blackley first announced a causal link between pollen and the disease in 1873.^ Morell Mackenzie, a British physician in the 1880s, cleverly pointed out that "summer sneezing goes hand-in-hand with culture, we may, perhaps infer that the higher we rise in the intellectual scale, the more is the tendency developed."^ By the late 19th century, then, hay fever had moved on from the more extreme ends of the aristocracy and now marked out bookishness.

Then, in 1910, a novel was published that made sense of these symptoms and their absurd, incongruous and ironic relationship to modernity.

E M Forster's *Howards End* is a novel about the tensions that existed between the country and city. They are all in there: the newly-monied classes who are the first-adopters of the motor car and have to do callisthenic exercises to stay in shape. These industrious people all possess bodies allergic to the very countryside they have tried to make their home. The new working class of sedentary workers in the book also have bodies deteriorating due to pathogenic labour and diet. The two sisters at the centre of the story become the natural ascendants to the house of the title. As if to proclaim this inheritance, at the novel's close their immunity to the disease and to the countryside is heralded by a staunch and final exclamation point. "The field's cut!" Helen cried excitedly, "the big meadow! We've seen to the very end, and it'll be such a crop of hay as never!"^

Even as this scene plays out, there is the "red rust" on the horizon, because "London's creeping"; the begrimed metropolitan monster is tramping slowly and inevitably across the Downs to consume them all.

That new ways of living were destroying the old was a commonplace idea throughout this period; but the notion that new ways of living were making it increasingly impossible for some to retreat to a simpler life was not. This was something new.

At the turn of the 20th century, the monied classes found themselves trapped. Born into an environment their bodies did not recognize, they developed a sometimes fatal immune defence against the very mechanism that in their early lives would have prevented them developing hay fever or asthma symptoms at all. How those allergies went on to develop into chronic and fatal diseases affecting billions is a story for another century.

It was only a matter of decades before that the scales had first tipped toward urban populations outnumbering rural ones in 1851, but already the human body was responding. Something had begun to go very wrong with *Homo*. A sensitivity to pollen and grass is in no way a favourable trait for any human on the savannah with its grasses forever rippling like the tide. But this new allergy made it impossible for some to put down roots in a place where they could breathe with ease outside of the industrial landscape.

With the 19th century descending below the horizon, the Anthropocene body found itself incompatible with the fast-growing urban world, and yet it could not return to the grasslands that had been its cradle and sustenance for millions of years.

WINDING BACK

1 Brisk walking and running are good for spine and knee health

It is a common misconception that running is bad for your knees; it is not – being sedentary is.

For a long time, walking and running have been known to contribute to bone density, but recent research is now showing that they are good for intervertebral disc health, too. Any kind of movement or exercise that also includes some dynamic stretching is good, too. The utility of static stretching is yet to be proven effective, but the dynamic kind increases mobility and range of movement: two things easily lost with Anthropocene habits.

2 Get apped up

Try anything that helps to motivate you out of the door. There are a number of good Couch to 5K apps: the Active 10 app (which came out of Public Health England's research into inactivity) is a good choice, too. Those Victorian clerks who walked miles to and from work every day had, even with a poorer diet, less osteoporosis and arthritis than sedentary workers today.

3 Learn to squat

This is not easy. As an Anthropocene human you will have spent a good deal of your life training yourself out of being able to do this by sitting in chairs for ten hours a day. But we used to rest, eat and converse like this (indeed, many people still do).

Start off by trying to squat with some support – a TRX at the gym, perhaps, or good sturdy table. Drop down, throwing your weight deep into your hips. Knees should track over the feet (not inside or outside). The stretch is likely to be felt in the calf muscles, around the groin and lower back. Try to relax into the position. It will also be easier if you have some sort of heel-raise – a rolled-up towel, for example. It is better to get into a comfortable position that you can stay in with your heels raised than to hunch over just to get your feet flat on the floor. Ideally the chest should face forward rather than hunching over. In time (and I mean months, not days) you will be able to do it without shoes, and it will be comfortable for reasonably extended periods. It is a great way of opening up your hips and lower back.

4 Avoid hibernation

Unless you have to sit in a cinema or theatre where you are confined, try to avoid developing habits that normalize sitting for extended periods. Take regular breaks and think of ways in which you can naturally interrupt sitting time without it being a chore to do so. Introduce some artificial habits that are not focused on minimizing effort (go out for a coffee instead of making one at home, don't park so close to your place of work, and so on).

Think of ways that will allow you to introduce more movement into your day. This can be tricky, as the default mode of the Anthropocene human is to do things as quickly and easily as possible (time is money), so a conscious effort is required. Try to use transport a little less. Don't allow time between meetings and appointments, *make time* to move between them. They will almost certainly go better if you arrive with a clearer head.

5 Most important of all: walk!

This is the single easiest way of introducing the kinds of movement that keep Sardinian shepherds on their feet well into their 70s, 80s and 90s.

6 Tell them to go out and play...

Professor Seang Mei Saw, Professor of Epidemiology at the National University of Singapore and Head of the Myopia Unit at the Singapore Eye Research Institute, told me that current research suggests that children need to have access to daylight for approximately three hours a day for healthy eye development. In children, reading two or more books per week is a good predictor of shortsightedness. It is important to remember, too, that dealing with myopia is not merely a matter of buying and wearing some glasses (for hundreds of millions of people in lower-income nations this is not possible). Myopia can also be a gateway condition that can go on to cause high myopia as well as more serious complaints in older age that can lead to blindness.

There are over a hundred good health reasons for getting your children to go out and play; significantly lowering the chances of shortsightedness developing is just one of them.

7 Get houseplants...

Houseplants are known to have a beneficial effect on indoor air quality (with the added bonus of having a measured impact on our general psychology, too). Plants can absorb some noxious gases (especially benzene, carbon monoxide and formaldehyde) and recycle them for us. The vapours emitted by soil bacteria also increase serotonin levels (the neurotransmitter that affects mood, among other things). Research has shown that hospital patients recover more quickly in rooms with plants. They humidify the air (so can help prevent colds and sore throats).

Best of all, if you are a little shy with the feather duster, they remove dust from the atmosphere. The slight negative charge of the leaves works by attracting airborne dust particles. Much of the dust will be absorbed by the plant, with some accumulating on its leaves over time. Peace lilies, *Howea forsteriana*, fiddle-leaf figs, and ficus can all be taken into the shower for a quick rinse when their leaves are dusty. With houseplants installed, the dust levels in a home can drop by up to 40 percent. This is particularly good news for allergy sufferers who are often affected by dust or the excrement of dust mites: less dust means fewer dust mites (don't Google how many are living in your mattress; it is better not to know). As well as this, the plants break down mould spores which can have a detrimental effect on people with lung complaints.

In 1989, NASA began a trial to discover which plants could clean the air on space stations with the greatest efficiency, in partnership with the Associated Landscape Contractors of America. The plants that made it on to their list have the capacity to clean at least 9 square metres (100 square feet) of indoor space (home or office). Plants such as English ivy, peace lily, broadleaf lady palm, Boston fern, the rubber plant and the spider plant are all excellent air cleaners. One of the best is the variegated snake plant (Sansevieria), which, if you're no good with houseplants, is practically indestructible and has the added quirk of producing oxygen at night.

New or novice plant owners may plump for a small plant out of caution. This can be a mistake because they dehydrate more easily; small plants can be difficult and sensitive charges. A medium-sized or, better, a larger plant will need much less attention and watering, and will process more air in the room. Knowing when to water is important. In most cases, it is a matter of sticking your finger an inch below the level of the soil, and if it feels dry, water it – if not, leave it. Always let the excess water drain away so the roots don't drown.

Humans have nurtured houseplants for thousands of years. After Vesuvius blew, preserving Pompeii in volcanic ash, the remains of houseplants were found in dwellings among the corpses.

Getting houseplants for your home turns your living room into an air farm.

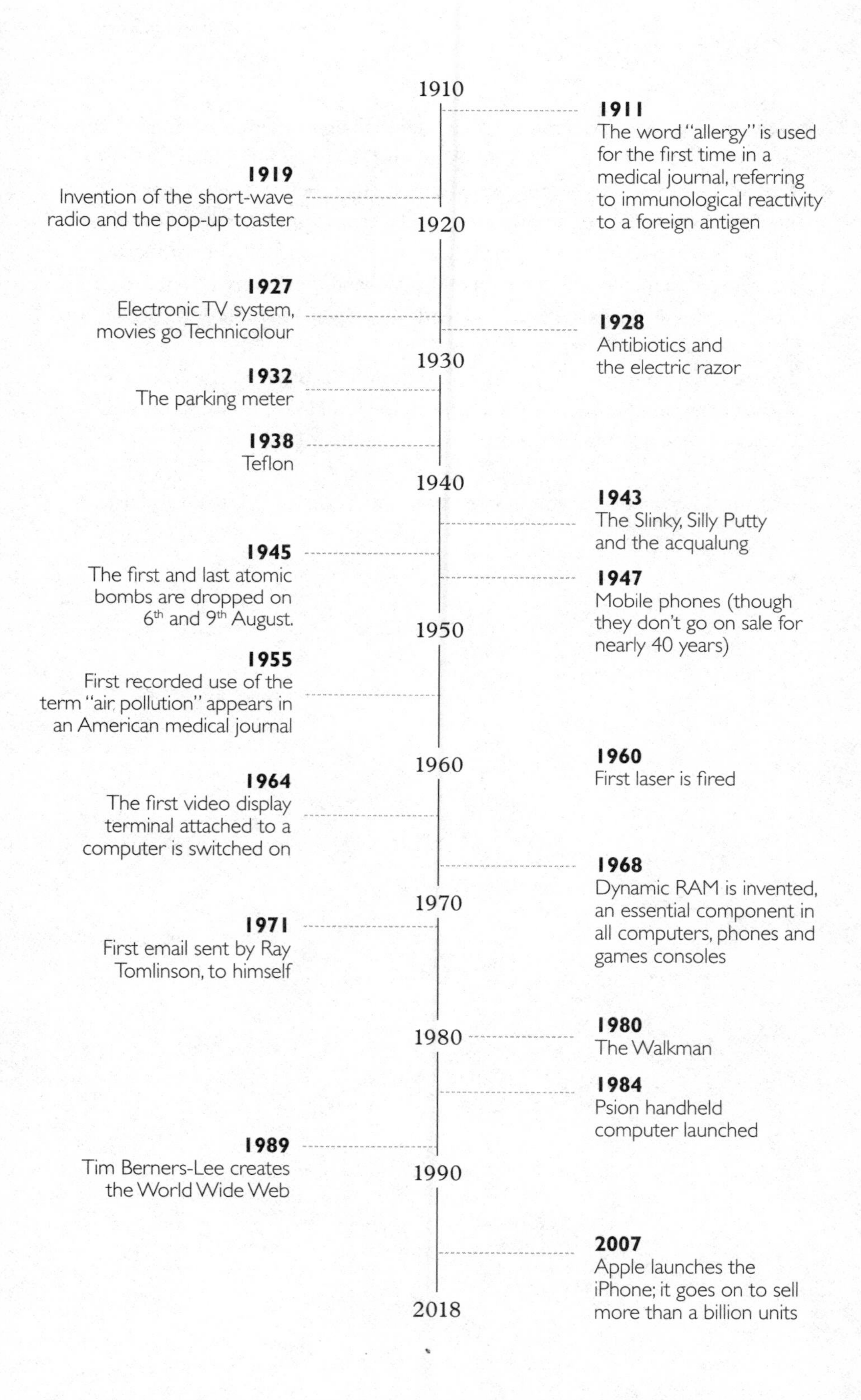
1910
1911
The word "allergy" is used for the first time in a medical journal, referring to immunological reactivity to a foreign antigen
1919
Invention of the short-wave radio and the pop-up toaster
1920
1927
Electronic TV system, movies go Technicolour
1928
Antibiotics and the electric razor
1930
1932
The parking meter
1938
Teflon
1940
1943
The Slinky, Silly Putty and the acqualung
1945
The first and last atomic bombs are dropped on 6th and 9th August.
1947
Mobile phones (though they don't go on sale for nearly 40 years)
1950
1955
First recorded use of the term "air pollution" appears in an American medical journal
1960
1960
First laser is fired
1964
The first video display terminal attached to a computer is switched on
1968
Dynamic RAM is invented, an essential component in all computers, phones and games consoles
1970
1971
First email sent by Ray Tomlinson, to himself
1980
1980
The Walkman
1984
Psion handheld computer launched
1989
Tim Berners-Lee creates the World Wide Web
1990
2007
Apple launches the iPhone; it goes on to sell more than a billion units
2018

Part IV

1910–present

THE SEDENTARY OR "DIGITAL" REVOLUTION

> The typist home at teatime, clears her breakfast, lights
> Her stove, and lays out food in tins.
>
> T S Eliot, *The Waste Land*

> … no material comfort can equal the luxury of a well-fitting, broad-toed, flexible, heelless shoe. Of course, the secret is that a good barefoot shoe enables us to walk naturally and to find in simple natural exercises not only health, but sanity and happiness as well. If I were a fairy and asked to bestow one gift on the man and woman of the twentieth century I would give them each a pair of model shoes.
>
> Bliss Carman (Canadian poet), 1908

The factories did not disappear at the end of the 19th century. Manufacturing continued to dominate for at least the first half of the 20th century. The sustained rise of very different kinds of industry in the late 19th century, such as banking and finance, insurance and accountancy, gave rise to neologisms to describe their difference from manual and factory labour. "Pen-pusher" was first recorded in the *Oxford English Dictionary* in 1875, with "paper-pusher" appearing during the Second World War.

Throughout the 20th century came the literal and metaphorical rise of the office block. And with it came another change in the ecology of labour and increasing refinement of modes of work and the ways in which it can be performed. What happened during this period is that the nature of labour for the Anthropocene human became increasingly simplified as our technology became smarter and more efficient; we exchanged biological calorific expenditure with mechanical energy derived from fossil fuels; but we gave no notice to our bodies.

Even though there is ecological variety in this economy it no longer expresses itself physically. In office blocks, people may perform a whole variety of different functions, but they do exactly the same kinds of physical work. Lawyers, accountants, bankers, academics, management consultants, administrators of all kinds in all industries do exactly the same kind of work that involves sitting at desks or computers. Board members assembled in a meeting for an international bank making multi-billion-dollar decisions do exactly the same kinds of physical work as students do in a lecture theatre.

The Metropolitan Revolution – the rise of cities from about 4,000 years ago to the present – changed numerous aspects of our political, social and daily lives, but made less significant alterations to our appearance. But the Industrial Revolution changed the ways in which we worked and the kinds of things we produced as well as the food we eat. It also hiked our number – once the machines arrived, populations which had been growing slowly for thousands of years began to spike, the atmospheric balance began to shift with greater levels of CO_2 and the balance between rural and metropolitan life began to shift.

We are now comfortably ensconced in the sedentary revolution, a mode of living in which work *and* life have become dominated by two things: the amount of time we spend working and our physical inactivity. The populations of developed nations are persistently nagged to keep active, despite being force-fed modes of labour, leisure and consumption which encourage just the opposite. We are not as far removed as we might like to think from the machine-fed Charlie Chaplin. But instead of being force-fed food, we are malnourished in an environment in which individual agency and action are required to negotiate and avoid sedentariness and where it is reduced to a personal responsibility rather than a completely natural way of life. Many government directives and schemes focusing on our levels of movement fail to understand that, and getting a dietician to wave a celery stick at us in the hope that we will change our behaviour is unlikely to be an effective solution.

Sitting is what we have been taught to do since infancy. From an early age, it is a social marker of good and appropriate conduct (and obedience, after all, could have been a matter of life and death in a factory). At school, our great-grandparents, grandparents and parents would all have been disciplined into sitting still for extended periods, in training for the sedentary life that lay in wait for them. As adults today we are encouraged to sit in our workplaces, at home, on public or private transport, in theatres, cinemas, restaurants, bars, meetings, classes, churches, social gatherings; the list goes on and on and on. The effects are widespread within our bodies in terms of our health, but also in our biomechanics.

Initially, trees dominated the landscapes around our dwellings. Then, during the Industrial Revolution, chimneys and their smoke became the defining symbol of economic success. In the post-industrial landscape, as the chimneys were demolished, even bigger office blocks sprang up in their place.

THE BIOMECHANICAL DOMINO EFFECT OF OFFICE LIFE

> Our bodies' highest order of intention is survival, so we don't heal to optimization, just to functionality. And if there's discomfort somewhere, it might be that the body has organized (or more commonly, disorganized) itself to manage a problem somewhere else. Its intention or strategy is to protect certain areas, so they're not moving, then other areas have to move more to compensate. And in a closed-chain environment, if one thing closes down and another has to open, you get imbalances. What I think is really important for the body is efficiency, effortlessness, and energy conservation, and those things tend to come with balance.
>
> Gary Ward, *Anatomy in Motion*

The rise of the office block heralded a new, modern way to work. People aspired to get out of factories and into offices, where the environment was cleaner, safer, and the work was – despite the protestations of almost any office worker today – easier. There was great fun to be had. Instead of having your limbs mangled to Bolognese by a machine your master forced you to climb inside to clean while it thrashed away at full speed, you could photocopy your bum cheeks, make a giant ball out of elastic bands, customize your desk with day-glo-haired gonks and pictures of your spouse. If there is work in heaven, surely it would be this?

However, in order to have office blocks you need big companies, and these didn't exist in the medieval period. It was not until the 18th century that there were ballooning organizations such as the East India Company, the first commercial endeavour to employ thousands of workers to fulfil administrative tasks. Following in their footsteps in the 19th century were the railways, banks, the post office, insurance and big retailers.

Toward the end of the 19th century, as the second wave of a technological revolution gathered pace with inventions such as the typewriter, telegraphy and the expanding uses and applications of electricity, the labour market also began to change.

The new office clerks were the fastest-growing occupational group in the latter half of the 19th century. In 1861, the census suggests that about 91,000 people were performing administrative work. By 1891, they had more than quadrupled, numbering just under 400,000.^

Well into the 20th century, manufacturing was still the main player in the economy. The de-industrial revolution crept in over a few decades and by the 1970s was in full sway.

Change is often easier on those holding the reins than on the people who have to trade in their knowledge and expertise only to find the exchange rate has fallen unfavourably against them. Despite the three-day week and the rubbish piling up in the streets because of strikes, there was a great deal of optimism in the 1970s about the future of consumer culture. Automation heralded a new age of leisure, certainly, but also of work. Economists scratched their heads wondering how we in the 21st century would fill all our spare time once much of our work had been offloaded to robots. (At least half of that notion is still true, though now we are substantially less optimistic that the automated future will lead to greater freedom. The 20th century forgot to ask an important question: who will own the robots?) Our material environment shapes us, but it will not save us.

In the offices of the 1970s, the photocopier arrived, followed later by floppy disks, laptops,[1] laser printers, voice recognition – and of course the PC. They were all beige, and they all made their operators more static. It was an anatomical revolution. Just as the Industrial Revolution permanently changed the bodies of its workforce, sedentary office work has done this, too. If the variety of work done was simplified in the 19th century, by the end of the

1 Though whose lap they were supposed to fit on is unclear given their original size.

20th, the process of homogenization was nearly complete.

This chapter will look at a selection of behaviours and illnesses that have been formed through, or associated with, office work: what sitting for extended periods does to the body, what it does to our feet and why nearly all office workers seem to struggle with yet more and more persistent back pain.

THE TALE OF THE TILTING PELVIS

Ergonomics has become increasingly important since its reintroduction in the mid-19th century. As we might expect of any science, it has improved a great deal in the last 50 years, and so has its advice on how to construct and design a workstation.

Current guidance is that we should stop working every 20 minutes for a 1–2-minute stretch break and that every 50 minutes we stop for 5–10 minutes and execute a completely different task. Our bosses would choke on their muesli if they saw us doing this. It is obviously very good advice (it comes from one of the leading ergonomics research centres at UCLA), but the cultural shift that would be required to make this an acceptable practice would be substantial – it seems unproductive.

If all workers had done this throughout their careers, how many would be reporting back, neck, shoulder or wrist pain? How many would be able to stand up correctly?

Sitting down for sustained periods is known to be bad for the human body in lots of ways. Being still for sustained periods impedes blood flow, which means metabolic waste products cannot be drained from the muscles. One of the other common beliefs about sitting down for extended periods is the damage it does to a set of muscles called the hip flexors.

The hip flexors flex the hip (just as a bicep flexes the arm). They also rotate and flex the spine; they adduct the leg (the latter half of the sideways karate kick), internally rotate the thigh and flex it, too. You can't really touch your own hip flexors because they run and connect to various points between the legs, pelvis and spine. One of the more significant pairs of these muscles is the *iliopsoas*. This muscle starts just below the hip joint at the top of the leg, runs through the pelvis and then fans and connects to several vertebrae. It is attached to the last thoracic vertebrae (the rib), and also connects to all the villains in the back pain saga (lumbar spinal discs L1 to L5); it runs over the ball and socket of the hip joint to attach at the top and inside plane of the

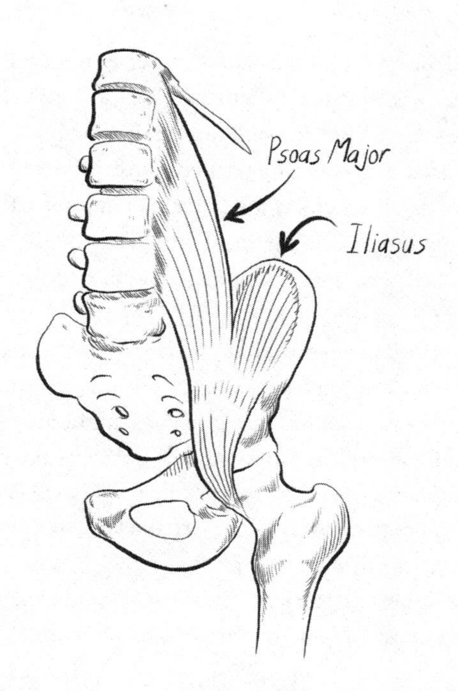

The two largest hip-flexing muscles – note their "depth" in the body, away from the skin surface. Together, they are known as the *iliopsoas*.

shaft of the thigh bone. These are the muscles that allow you to lift your leg up toward your chest. They are key for any kind of walking, running or jumping.

Davis's Law says that soft tissues adapt to demand, which suggests that the hip flexors eventually respond to extended sitting by shortening in length.

The *iliopsoas* (along with the other hip flexors) are extended when standing and relaxed when seated. Their mechanism is not easy to describe, but while sitting, imagine wrapping an elastic yoga band around the back of your knees and then tying it behind your neck. The band would be long enough for you to be comfortable while sitting with the band in place. If the band remained in position, when you tried to stand up, one or both of the following would happen: there would be considerable pressure on your neck (where the band is tied), and you would not be able to stand up fully. The *iliopsoas* operates in a similar manner, but the mechanism that they attach to is the lower back. If the structure tightens it can lead to excessive lordosis (inward curving) of the spine.

The mechanics of this, to me, make perfect sense, and there is a reason why 54 percent of people who report lower back pain in the US are sedentary workers. It is a good hypothesis. After all, don't women who habitually wear high heels find it difficult to impossible to transfer into flats? In heels, the angle at the ankle is quite obtuse, whereas in flats it is perpendicular. The discomfort derives from the fact that their calf muscles have adaptively shortened and can no longer stretch comfortably to a resting position.

Many sedentary workers also have something called "anterior pelvic tilt". This happens when the pelvis visibly tips forward. Imagine Channing Tatum (it could be anyone, but why not him) standing up, viewed from the side. If he was wearing a belt, we would see that it was practically parallel to the floor; he is in good shape, Channing.

When viewed from the same aspect, a sedentary worker's belt line would probably be tilting downward toward the front. This condition is thought to be caused by the shortening of the hip flexors (as well as the lengthening of the hip extensors), but also by weakness in the surrounding muscles such as the hamstrings, glutes and stomach muscles. The pelvis's strategy is to tilt away in front of us, and the mechanism then overloaded when trying to maintain the pelvis is the discs and vertebrae of the lower back.

The changing shape of the spine and the associated muscular imbalances are often caused by prolonged periods of sitting. A lack of stretching or

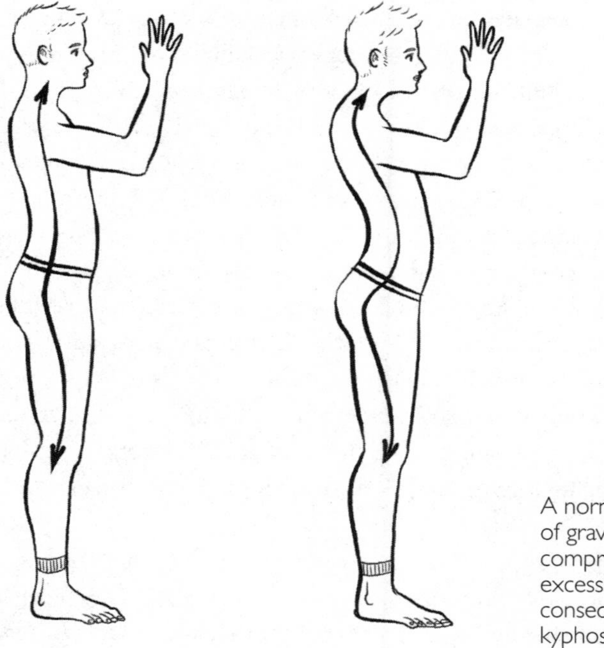

A normal centre of gravity (left); one compromised by excessive lordosis and consequent corrective kyphosis (right).

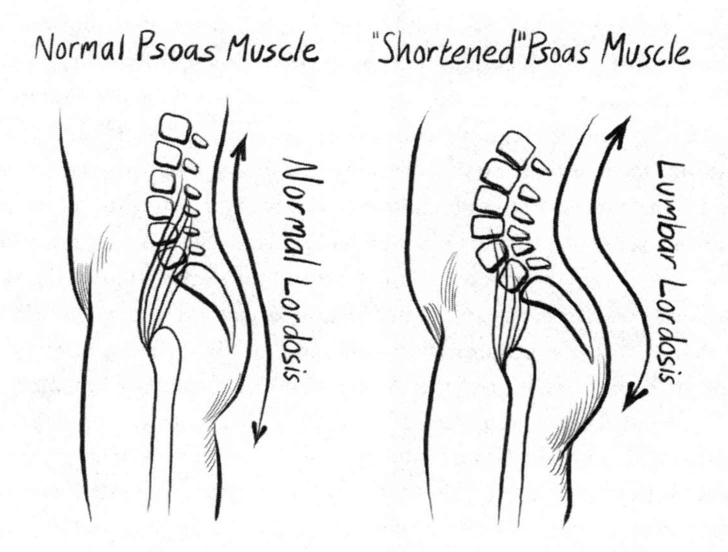

How shortening of the hip flexors might increase lordosis (though unlikely).

strengthening exercises also contributes to anterior pelvic tilt.

The symptoms are there, but what about the diagnosis?

For several months, I was convinced I had found the source of my own back pain because I mostly did two things; I sat and I ran, and all the literature told me that these were muscle-shortening activities. And with a weak core, I was not able to support myself while running. It made sense. So I stretched the hell out of my hips, planked and crunched my way to a stronger core and what was the result? Not much, except for the fact that I could plank and crunch more than I used to. And my back still hurt.

What is "the core"? What ideas does that word conjure? For me it sounds like something as substantial as a diesel engine, strong and tough. But how much muscle do we really want there? In the place that we think of as an area that should be densely filled with tough muscle, there are about 8.5m (28ft) of intestine, tens of trillions of microorganisms, reproductive organs (women grow babies in there) – and around these are more supportive soft tissues and bone. Should this really be called a *core*? Or might it be more helpful to think of it as a set of muscles, tendons, bones and ligaments that have to work together performing functions as complex as that of any of the soft tissues?

The idea that we should tense our core muscles during activity is a prevalent one. But the reason we have muscles connected to *any* joint is to promote movement. If you try to lift your pelvis as you walk, you tense your abdominal muscles, and in doing so you lose the counter-rotation in your pelvis that you need to complete the gait cycle. The pelvis and the ribcage are a little like the rear foot and the fore foot; on a lateral plane they are supposed to move in opposition to one another. Counter-rotation cannot take place because a tensed core allows no movement between the ribcage and the pelvis. Core strength and core stiffness are not the same thing, and the latter is probably worse for the spine than sitting for extended periods. What we need is not stiffness, but mobility, this being the ability of the core to react to the appropriate motion in the structures which encourage locomotion.

To my mind there are three quite substantial objections to the *iliopsoas*-shortening theory. The first is that despite its prevalence as an idea (there are lots of diagrams and articles about it on personal trainer websites) there is no science to support it. Without science, it becomes "circumstantial evidence".

The second objection is that Davis's Law tells us that tissues adapt to demand, and anyone who has performed any kind of yoga or stretching regime can tell you that a few minutes every day for a few weeks will make a huge difference to a given muscle's length and strength. Why should the *iliopsoas* be any different? The spine is a robust structure, with more muscles and tendon attachments than anywhere in the body, so are we to believe that over time it gives way, conceding to a muscle that is about the same thickness as a chicken breast? If the muscle shortened, it would just stretch when we stood up.

Finally, if the *iliopsoas* shortens as a result of sitting at a work station, why don't our biceps? The range of motion is almost identical. When we stand up from our desks, we have no trouble straightening our arms; why should the *iliopsoas* be any different?

This is not to say that nothing is wrong. Anterior pelvic tilt is common in our society; you can see it on any high street, no matter the time of day or how populated – it is that common. My feeling is that it is more to do with not using the surrounding musculature associated with sitting, rather than some pathological contraction of muscles that seem not to respond to regular stretching throughout the day.

It is all very well to tell people that they have no problems with their hip flexors, but what use is this if their pain persists after stretching and strengthening, which it almost always does?[2]

The sedentary work that first began its onslaught upon the bodies of office workers who believed they were sailing into a bright future of machine automation in the 1960s and 1970s is having a damaging effect on us. Some of this is no doubt caused by anterior pelvic tilt and lordosis in the lower spine, but the sedentariness is the real villain of the piece. It's not that we are sitting down. Neither is it the way that we are sitting down (there is no good or correct position to sit still in for extended periods). Instead, it is the fact that we are sitting still that is having the greatest impact on global health.

An anterior tilt to your pelvis is not a guaranteed signal of biomechanical disorder that will lead to chronic pain, but it is a likely sign that the pelvis has become more limited in its range of movement. Anterior pelvic tilt, just like posterior pelvic tilt, is a normal part of the human gait cycle and it looks as though sitting for extended periods impinges upon our ability to do this effectively.

Sedentary work and sitting probably unsettle the normal vocabulary of movement that the body expects to use on any given day. Sitting is probably weakening and deskilling the human body rather than shortening its muscles. Movement becomes like a first language that is rarely spoken, so that when you go home, the language and expression that was entirely natural to you has become strange and limited through lack of use. The beginning and end of the sentence is clear in your head, but how you navigate from the first part to the last has, oddly, become a mystery.

MAKING A STAND (AGAINST SITTING)

In recent years sedentary behaviours, and sitting more specifically, have come under increasing scientific scrutiny. The research is stacking up that shows it is *significantly* associated with several of the major killers. A study published in the *British Journal of Sports Medicine* in 2014 found the following:

2 It is really hard to explain chronic pain to someone that doesn't suffer from it – Elaine Scarry said of pain that, "Whatever pain achieves, it achieves in part through its unsharability, and it ensures this unsharability through its resistance to language." (Scarry, *The Body in Pain: The Making and Unmaking of the World*, Oxford University Press, 1985.) And while the pain is unpleasant, it is intensified by the constant nagging belief that comes with every shard of pain, that you are becoming more disabled, that your back is crumbling; that you are trapped in a process of being broken and breaking at the same time.

> Sitting time was significantly associated with adverse levels of waist circumference, body mass index, triglycerides, HDL-C [the *good* cholesterol], insulin, HOMA-IR, HOMA-% B and 2 h postload glucose [means of measuring β-cell function, insulin resistance and fasting blood-sugar levels], but not with blood pressure or glucose level. In stratified analyses, sitting time was most consistently related to cardiometabolic risk factors among low and middle socioeconomic groups and for those who reported no weekly physical activity, but there were few differences between sex or race groups.^

This is all very bad news for sitters.

Another study, published in the *American Journal of Epidemiology* in 2010, found that "several factors could explain the positive association between time spent sitting and higher all-cause death rates".^

This is confirmed in the *British Journal of Sports Medicine* study (mentioned above), which had concluded that "self-reported sitting time [among 4,560 participants] was associated with adverse cardiometabolic risk factors consistently across sex and race groups in a representative US sample, independent of other risk factors. Excessive sitting warrants a public health concern."

In the last four decades in the US, the number of people suffering from type 2 diabetes (to which sedentariness is strongly linked) has risen from below four million to well over twenty million: that is nearly 10 percent of the population. After only a couple of hours reclining our bodies go into a sort of hibernation, which sounds fine – hibernation is pretty natural isn't it? Perhaps if you're a skunk or a groundhog, then yes. Hibernation is what animals do to survive when food sources are life-threateningly scarce. Hibernating animals drastically lower their metabolism to store energy as fat.

Every two hours we spend sitting on our portable carry-cushions reduces blood flow, lowering blood sugar and increasing the risk of diabetes, obesity and heart disease. The result for humans is that even when adults meet the published guidelines for daily exercise, they can still be at risk. A study published in the journal *Diabetes* in 2007 suggested that sitting for long periods, despite a burst of exercise, can still compromise metabolic health.^ This was echoed in a 2010 study at the University of Queensland's School of Occupational Health^ and many others. Although we garner some of the beneficial effects of exercise, it is not sufficient to offset the damage done during long sedentary periods; this is a little bit like playing football regularly and smoking 20 cigarettes a day.

What the research seems to be saying is that we need to be active either at regular intervals or throughout our day – whether at school or at work. There is a widely held misunderstanding that sitting for too long is the same thing as not exercising; the two are quite separate things. You can exercise and still be classed as sedentary because of the sum of the hours you spend sitting.

The boardroom in any company is where the money is spent. As at King Arthur's Camelot, the big table boasts power, wealth and success – and being invited to sit at it is like being made a knight in your company or institution. And the chairs… even though they are rarely sat on, hundreds are spent on each one. The whole set-up is a symbol of the important decisions that are made in that place and around that table. But they mean early death for those who regularly sit through these meetings. It is time we retired them before they do further damage. We certainly do not need to be working out all day; we do not need to hold meetings in the gym or during Boxercise classes, but we do need to have meetings while on our feet, not seated. A rather pleasant additional benefit of more vertical meetings is that they become much, much shorter and more efficient.

Perhaps the real problem is that people don't see sitting as a disease. Technically, this is correct, but sitting is a major cause of a sackful of diseases. The number of smokers in the West has plummeted in the last couple of decades, and the single most likely reason for this is that the populace is now better educated about its associated health risks. We wouldn't expect that just because someone is thin they are unlikely to be made ill by smoking, and the same goes for inactivity.

A 2009 study at the University of South Carolina concluded that "normal-weight adults who are sedentary are at increased risk for cardiovascular disease-related outcomes than are overweight or obese adults who are aerobically fit", suggesting that it is healthier to be fit than it is to be thin.^

There are dangers associated with sitting, and if people are better informed about these risks, they might be in a better position to prevent some of the millions of associated deaths.

Sitting disease is a silent killer and the results from some of the research are alarming. The 2010 American Cancer Society study looked at the amount of time subjects spent sitting compared with the levels and types of physical activity and produced some mind-boggling results. The women involved in the study who were inactive (sedentary for more than six hours a day) had a

94 percent higher risk of *dying during the duration of the trial* than those who were sedentary for less than three hours a day.^

A 2012 study published in the *American Journal of Epidemiology* drew links between sitting and premature ageing^. The researchers examined a telomere, which is a kind of protective cap on a chromosome that gradually becomes thinner with age and so is an indicator of cellular ageing. They discovered that those of the 7,813 subjects who were sedentary for ten hours a day had significantly reduced telomere length, which aged them biologically by about eight years. The study concluded that "even moderate amounts of activity may be associated with longer telomeres".

A small part of the research I conducted for this chapter was traditionally seated in a chair. But I very quickly swapped to reading an iPad while walking on a treadmill when I discovered that those in the 45–64 age bracket doing sedentary work (ie me) were 40 percent more likely to enter a nursing home during retirement.

Today's service/knowledge workers, the new working class, are unknowingly exposing themselves to an environment that is noxious and unsafe and is changing the ways in which their DNA is expressing itself, leading to all kinds of pathologies and illnesses.

Our environment has changed significantly and in so many ways. The number of people who worked in the knowledge economy during the earlier phases of the Agricultural Revolution was of course zero. In the 18th and 19th centuries, the number began to crawl up very slowly. At the beginning of the 20th century only about 10 percent of the workforce in the UK worked in the knowledge/service economy; now it is closer to 80 percent. And exercise alone is not the answer; not for adults, and nor for those children that we are teaching to sit still.

Teachers who take their kids outside for a more free-range form of education are seen either as creative and innovative, or naive and indulgent, but there is no reason why children have to sit at school; while I'm not sure that treadmill desks are the answer either, there are numerous solutions available to us.

Some schools in the US Midwest trialled the use of stationary-bike desks; they are more common than you might imagine. With poor ergonomics, they are unlikely to catch on unless they are used for the purposes of passive learning, such as reading. A few years ago, a grade school in North Carolina trialled the use of "read ride" time and found it produced some surprising results. At the end of the academic year, they discovered that the students who had spent the majority of their time in the program necked ahead by 42 percent in reading proficiency compared with those who hadn't (do note, the results are not scientific as the groups were self-selecting, but a proper trial would be worthwhile). There were other benefits to the program, too; it burned calories non-competitively; overweight students did not have to worry about performing badly or being picked last, no one's distance was measured and students could take a break whenever they pleased.

Standing-biased desks might be a viable solution, too. They burn more calories throughout the day, and have the added bonus of increasing the attention span of the student (or indeed worker). A trial conducted with 480 schoolchildren and 25 teachers examined the introduction of these desks, which encourage students to stand, though they have a stool to sit on if they want a rest. Mark Benden, who works in Environmental and Occupational Health at Texas A&M University, ran the trial and reported that follow-up interviews with the teachers revealed that the children were able to focus for much longer periods, and that younger children were less resistant to standing than older ones.^ Perhaps traditional methods had begun to succeed with the children that had already been through several years of schooling.

A BACK FOR THE FUTURE

What problems might early sedentary behaviour be storing up for the future? Going straight from school into office or call-centre work might be a bit riskier than it first appears. Anyone reading this will already be aware that back pain is a global problem. During the 19th century, when manual labourers were injured, it was usually a dramatic event, a sudden trauma. Today back pain differs in that it is both more chronic and more persistent. While reporting back pain is almost certainly on the increase (compared with other centuries) the reported numbers of back-pain sufferers are also confusing.

Some nations in the recent past reported very little back pain (places such as Bolivia, India, Mexico, Portugal and Vietnam), but these countries are fast catching up with the global trend in which most countries rank it as the number one disability.

In the US, back pain is thought to cost in the region of $100 billion annually. The numbers imply that back pain is worse in higher-income nations than in low, but there is a lot of statistical noise in them. Almost every country in the world reports an increase in the impact of back pain, but what is rising? Is it pain, the cost of healthcare, the reporting of pain, injury, loss of working days? These are different things with differing determinants. The numbers are perplexing and the science isn't much clearer.

The Victorians had no occupational training for heavy manual work. There was no one to show them the basics of lifting technique. We do have training today, but it doesn't seem much help. A massive review which collated the data from 1,827 previous trials concluded that "manual handling training is largely ineffective in reducing back pain and back injury. High priority should be given to developing and evaluating multidimensional interventions, incorporating exercise training to promote strength and flexibility, which are tailored to the industrial sector."^

If lifting technique is not to blame, what is?

What we can tell from our Anthropocene journey thus far is that back pain may have started with sedentary behaviours in the 19th century, but since office work became the norm, it has spread like a freshly cracked egg hitting a frying pan.

More than a million back injuries occur in workplaces each year in the US alone. The group that reports the highest rates of injury are truck drivers, who have the most toxic mix of long periods of inactivity punctuated by heavy manual labour. Other high-risk professions for back injury are nurses, labourers and cleaners. What is surprising is that these are all active professions, and on first glance seem like they should be healthier than office work. Evidence suggests, though, that people can take a few years to adjust to an occupation – for example, back pain in nurses is most frequently experienced by trainees (this was probably the case with 19th-century dock workers, too). Cleaning is infrequently a lifelong career, and often employs those ill-used to such high levels of activity. Truck drivers intersperse heavy lifting with long bouts of sedentariness. The solution seems clear; if you want to prevent a back injury, stay at home.

But those doing nothing seem to fare even worse.

Low back pain is eight percent more prevalent in high income areas of North America than in low and middle income ones. More than half of those who experience low back pain spend their working day seated. The luxurious life is not going to give your back what it needs.

What is so surprising is that back pain is persistently mysterious. While experts are confident that pain is often muscular, with disc and vertebrae pathologies coming joint second, the difficulty arises in deciding which of these structures is the culprit in any given person.

Professor Mike Adams from the University of Bristol has been working on back pain since the 1970s and there are few people who share the range and depth of experience that he has. He explains that discs in the spine become damaged in essentially two ways: through injury or fatigue (where loading outpaces the ability of the structure to repair itself).

One of the key problems is the complexity of the configurations in and around the spine. An injured muscle can take a matter of days to heal because it is a single structure that is very metabolically active. Bones and tendons (with a lower supply of metabolites) can take weeks or maybe a couple of months to heal. There is a lot of cartilage on the spine; damage can take years to repair. But Adams says, "the collagen turnover time in discs is estimated at two hundred years." Because a disc is a comparably self-contained unit, with only limited access to blood supply, it cannot heal itself very efficiently. Over time, damage can easily outstrip repair and fatigue accrues.

This is why it is so important for young people to remain active: it increases tolerance of fatigue and injury.

Muscle power (which is strength multiplied by velocity) is essential to independent living in middle age and later years. Old-age immobility is a solid predictor for early death. It is easier to acquire strength during childhood and into early adulthood; after that it drops away.

This may not sound like a revelation, but it is interesting that the rate at which power drops away in our muscles is fairly consistent whether you are strong or not. Whether you are a coder or an Olympic weightlifter, the slow decline after reaching your peak is inexorable. This means that the slope in the graph from about the age of 30 is just as steep no matter what the starting point. So those who remained sedentary throughout the first half of their lives start from a lower peak and fall below the strength threshold sooner (the threshold being whether you can bathe or get off a toilet by yourself).

Adams explains, "As far as muscle is concerned, do it till it hurts. For bone, if you're young, go weight training and 'grunt'." He goes on to say that more caution is needed in middle age. "As you get into your forties, you don't want to load your spine too severely and relentlessly. Beyond the age of about 45, the number one consideration has to be not sporting performance, but the avoidance of injuries, especially to cartilage and discs."

Inactivity in the young, as well as in adults, stores up significant problems for the future.

Are standing desks the solution? They are selling by the truck load and with good reason. Research into their use in schools and offices endorses their benefits for both our psychology and physiology. Early in 2018, an article in the *European Journal of Preventive Cardiology* aimed to quantify how standing desks contributed to daily calorific expenditure. The answer was that 2.5kg (5.5lb) of weight could be lost in a year by transferring to a standing desk.^ But that is not the whole story.

For about half of people with chronic back pain the cause is likely to be some pathology with intervertebral discs, while the other half are most likely to have problems with apophyseal joints. They are the spine's stacking system, and just like the hip or the knee, they are lubricated synovial joints. The spine has a pair of these joints at each level, so there are more than 50 of them.

Adams explains: "90 percent of people over the age of 40 have osteoarthritis in one or more apophyseal joints. They tend to become osteoarthritic in joints with the highest stress, the lower lumbar spine and the neck. When apophyseal joints go wrong you get aching joints that stay sore for years." These joints are not designed to be load bearing, "they are there to stop you bending backwards too far, sideways too far, actually rotating too far, but their mechanical role in resisting compression is exaggerated when the discs between the vertebrae become narrowed. Walking fast flattens the lordosis and slightly increases the flexion in the spine, and that's enough to disengage the apophyseal joints."

Standing upright while hardly moving increases the lordosis (inward curve) of the lower back. This increases the loading at the outer edge of the vertebrae which can cause fatigue, and go on to become a cause of chronic pain. If you are one of the tens of millions of people (and I am) who have this kind of back pain, then a standing desk is actually likely to aggravate it.

A recent study from Curtin University in Australia has found that in a small trial of 20 people, the use of a standing desk caused "significantly" enhanced pain in the lower back and lower limbs.^ Many such desks have been provided by companies seemingly in fear of future legal action. There has not been much research into the standing desk, and for now it is certainly not a panacea.

How different, really, is a standing desk from the kinds of labour that early factory workers performed?

It makes more sense to have an adjustable desk that will permit you to vary your posture throughout the day: one you can stand at, sit at, raise up to lean on as you read, or that can be low enough for you to sit back in your chair and put your feet up. It is variety that is key. What is much more important, though, is to not have a job that requires you to be in one place for 40–50 hours a week.

Like so many other things, back pain seems to have been a baton that was handed on to us by the Industrial Revolution. But, if those alive at that time first documented their experience of it, we are the ones who democratized it. Back pain in Victorian histories seems the preserve of the genteel classes; we have made it readily available to all.

CORE STABILITY?

Along with MRSA, TB or HIV, is back pain a disease that spreads within healthcare systems?

Probably less than one percent of all back pain has really serious causes, such as a fracture, a cancer or a neurological disorder. About one-twentieth of sufferers have the identifiable symptoms of a prolapsed disc or similar. The remaining 94 percent of us (and we are talking about hundreds of millions of people) have no medically diagnosable condition. This is mainly because a precise diagnosis of back pain is so difficult; the structures are buried deeply at the centre of the body – there are well over a hundred ligament and tendon attachments along the spine; it is practically an ecosystem of muscle.

There is little or no science behind all the solutions offered – strengthening your core, an ergonomic intervention, a standing desk, learning to tense while going about everyday activities – in terms of curing back pain.

What is worse is that there are many costly but low-value treatments for back pain sufferers: MRI scans, physiotherapy, needles, painkillers all have proven limited impact on the condition.

There is also evidence that back pain may in fact be an iatrogenic illness: one made worse by the healthcare system. A study by Ivan Lin focused on a population that had previously seemed to be insulated from the effects of chronic low back pain: indigenous Australians.^ The authors explain, "despite suffering a tremendous burden of disease Aboriginal Australians have been uniquely identified as protected from the disabling effects of

CLBP [chronic low back pain] because of cultural beliefs. One study found that, despite a high prevalence of CLBP in one remote central Australian Aboriginal community, the impact of CLBP was small as few pain behaviours were observed and people did not seek healthcare."

The study found that most of the participants in the trial ascribed their pain to "structural/anatomical vulnerability of their spine. This belief was attributed to the advice from healthcare practitioners and the results of spinal radiological imaging. Negative causal beliefs and a pessimistic future outlook were more common among those who were more disabled. Conversely, those who were less disabled held more positive beliefs that did not originate from interactions with healthcare practitioners."

Back pain is not psychosomatic, but there is overwhelming evidence that stress increases pain signals and inactivity has robust associative links with back pain.

At the moment the evidence suggests that movement, physical activity and feeling positive about your strength and capacity are all more effective at treating back pain than any surgical or medical intervention.

Both yoga and Pilates encourage activities that are known to promote better pain management: they spur movement and the use of many joints and they reduce stress – just like a woodland walk. Each to their own. If yoga works for you do it because you like it; it's good for you. If it doesn't, try something else that has a similar impact.

Walking remains the miracle cure it always was. It's something that links us to those grassland species of millions of years ago and is good across the whole spectrum of being human. It decreases lordotic loading on the back and may stimulate disc health – this is really significant as a larger, healthier disc will help shield the apophyseal joints. And most important of all, it is not a sedentary activity. Everyone agrees that remaining static for extended periods of time is bad for us all.

Our backs come from the deep past, tracing their roots to more than 500 million years ago, and they are not ready for this future. They emerged slowly to become ideal in a specific landscape, and we have changed that environment almost wholly.

In the 19th century, although there was documented interest in pathologies of the spine and back pain by figures such as Edward Duffin, it was not a widespread global disability costing hundreds of billions in healthcare expenditure and lost revenue. There is no doubt it was there, but it was comparatively rare despite the fact that the work was harder and more physical than ours today. Back pain then had less opportunity to become iatrogenic as access to healthcare was limited. Beliefs about pain and disability were different, too. Many Victorians' bodies would have been stronger than those of sedentary workers today, implying that they may have had better stress shielding in their biomechanics than the Anthropocene human.

Today in low-income nations, labourers work harder than most Western factory workers, but their backs seem to adapt. The heaviest thing most sedentary workers lift today is their own upper body, which weighs about 30–50 kg (66–110lb).

If we are to develop backs for the future, we need to acknowledge that there is little opportunity for most of us to completely change the work we do without changing our careers. How many of us are able or willing to do that?

We can change the way we do our work to create stronger links with those grasslanders: it is they, after all, that our DNA still desperately wants us to be.

Back pain is the greatest cause of global disability, and the jury is still out as to whether the healthcare system exacerbates moderate symptoms and amplifies them to the point of becoming chronic, or a much changed working environment escalates the likelihood of disability – but it is most likely both. And in both cases, it is the environment that we have made that is still in the frame. It is no coincidence that the global spread of back pain is synonymous with the global increase in sedentary behaviours. The best solutions to back pain at the moment are not about special occupational training, standing desks, strap-on corsets or esoteric postural advice. Instead, what people need are spines that are accustomed to movement and regular use. As the office blocks began to spread in place of the factories, and chairs began to spawn like virile bacteria throughout the 20th century, opportunities for regular movement became as difficult to come by as a black sheep at a hyena convention.

IS OFFICE LIFE MAKING OUR FEET BIGGER?

As the office blocks went up and the decades fell, so did the arches of our feet. New experiments in living appeared with regularity as manual work gave way to sedentary office work. And exercise, which had for centuries been the preserve of the monied classes, from the ancient Greeks to the landed gentry in Britain, became something that affected the masses. The jogging revolution shambled into view in the 1960s and 1970s, then Jane Fonda got us high kicking and feeling the burn in our living rooms during the 1980s.

The office blocks that replaced the factories are certainly cleaner and safer, but hidden from the omnipresent strip lighting, the threat of morbidity still looms behind the filing cabinets. Ever since the cereal stalk mattresses of Ohalo II, we have revealed ourselves to be a comfort-loving species. That comfort is exemplified by a quick and unscientific experiment I've just conducted from the window of the Brighton coffee shop in which I'm sitting. Of the 20 people who have walked by, one was wearing boots, three were wearing shoes and 16 were wearing training shoes.

In the 19th century, running shoes were exceptionally rare. They were specialist equipment but could be found if you were willing to look hard for them. They were quite expensive, just as they are today, but the design was very different. They were light and flexible, without any impact protection – they could be crumpled up in your hand as easily as a sheet of A4 paper. They weighed about 28g (4oz) and resembled a cut-off leather sock. Toward the end of the 19th century some bright spark registered a patent for the use of compressed air inside cricket pads and the soles of shoes. But the patent did not take off. A few more companies got in on the idea of the sports shoe; Converse began in 1917, and still have the thin sole they started with, because it was not until the mid- to late 20th century that we started to see the heavily-cushioned shoes that went on to become trainers (such as the Nike Air).

In the 21st century, the training shoe industry has grown to be worth an estimated $55 billion (£41 billion) – about the same as the GDP of a small to medium-sized country. We are so beloved of them (I'm wearing a pair as I write) that market research has concluded that only 25 percent of buyers have any intention of using them for sport or athletics.

In the last 40 years they have also taken over the role of supporting us. Buying a pair of running shoes now is only marginally less complex than getting hold of someone to talk about your faulty broadband connection.

Shoes are designed for all sorts of practices and purposes. Are they for road, scree, trail, marathon? Are they a barefoot or racing shoe, motion-controlled, cushioned or neutral? Are they for people with flatter feet, or those whose arches might be "too high"? What about those with Goldilocks feet, where the arch is "just right"? Finding a shoe that does not support you is not all that easy. Research published in 2016 by biomechanics expert Hannah Rice at the University of Exeter showed that people who ran in expensive cushioned trainers were likely to increase their risk of injury. Rice's study found that "footwear alters the load rates during running", perhaps because buyers believe their new shoes insure them against injury.^ But surely support is good? Those boys and girls who worked the machines in the 19th century might have been glad of a little.

Outsourcing the support of our bodies is sometimes necessary. Professor Irene Davis from Harvard (with whom Rice previously worked), is a world-renowned expert on biomechanics and she uses the analogy of the neck brace. If you have an accident and hurt your neck, you might be required to wear a neck support for a while to limit and sustain your head movement while you heal. But imagine if you left the support on. What would happen? The muscles in your neck would atrophy. When you eventually took off the neck brace, you would almost certainly discover that you needed to wear a brace permanently. This is what support does for our feet. If we outsource the work of our feet to an external prosthetic, without the correct stimulation our feet weaken.

The foot sizes of humans have been relatively static over millennia, climbing and falling a little with both diet and height. But in the last century or so they have started getting bigger, growing by two sizes in the last 40 years alone. In the 1960s the average size for US women was 6.5 (a UK 4), today it's 8.5–9 (a UK 6.5).^ Our mean weight is a contributory factor, but it is not, surprisingly, just that; it is also because our arches are collapsing and flattening.[3]

A 2017 study of 500 Indian women and men found a mean rate of flat-footedness (*pes planus*) in about 14 percent. An older study from 1992 found that of 2,300 children aged 4–13, "the incidence among children who used footwear was 8.6 percent compared with 2.8 percent in those who did not."^

Another 2017 study found that "habitual footwear use has significant effects on foot-related outcomes in all age groups, such as a reduction in foot arch and hallux angles. The results indicate an impact of habitual footwear use on the development of the feet of children and adolescents. Therefore, growing up barefoot or shod may play an important role for childhood foot

---

3 The folklore about foot size is rich and often seems to have sexual connotations. However, a survey by the website Illicit Encounters in 2014 found that men with bigger feet (size 10 or over) are up to three times more likely to cheat on their partners.

development, implying long-term consequences for motor learning and health later in life."^

Today, estimates of flat-footedness in the US vary wildly, with some rates as high as 30 or even 50 percent (10 percent suffer with *plantar fasciitis* – a disorder in which some of the supportive tissue in the foot tears and results in heel pain). That's up to 150 million people who are snowman-melting into the ground.

What helps none of these figures is that our average weight is climbing, too. Lowered levels of activity are related to weight gain which in a feedback loop encourages lower levels of activity; both of which are bad for our feet.

If a house was collapsing, we would dig out the foundations and permanently replace them with stronger ones, but our bodies don't respond all that well to permanent assistance. Supportive shoes should be used like a neck brace or a plaster cast, to aid recovery in the short term. And part of that recovery should also be work done by the patient to restore better function to the intrinsic muscles, tendons and ligaments of the foot.

The same goes for the artificial heel raise in almost all our shoes. The average running shoe has a 15mm (0.5in) raise in it. A raised heel provides more cushioning and permits a longer stride, making it more comfortable to land heavily, and actually encourages us to land on the heel. These heel raises mean that we can walk faster and farther, too.

Because we are well protected, we can negotiate the environment without care. In shoes that have no protection, such as the leather boots worn in many medieval cultures (which were essentially thick socks) the foot strike would almost certainly have been what is called mid foot, with the heel and the ball of the foot landing at the same time. It is slower, but safer and probably accounts for the fact that Chaucer's pilgrims took so very long to walk the 80km (50 miles) from Southwark to Canterbury Cathedral in *The Canterbury Tales.*

The world we have made is easy to negotiate in appropriately cushioned shoes. The concrete epidemic of the last two centuries has made most public spaces either challenging or completely inaccessible while barefoot. The cultural taboo of bare feet is odd, too; for the sake of our bodies, we would be better off if we spent a lot more time without most of our shoes.

BODIES IN MOTION

So what can we do about our feet? And what does the state of our feet tell us about who we are and how we're coping? The easy answer to these questions is that there are no easy answers. Clearly, things are awry, but surely it is just a matter of performing a few strength exercises and stretches? While that is not a *bad* thing to do, it seems we need to know more about our feet, their movement and how important they are for our entire anatomy.

Gary Ward (a man whose knowledge I have cited already, *see* pages 50 and 190) is a movement specialist, anatomy theorist and pioneer of closed chain biomechanics (being a means of looking more closely at the connectedness of the body's mechanical structures). After a TV appearance he awoke one morning to find himself famous. Within a few hours of the programme going out his practice, Anatomy in Motion, had a 25-year waiting list after he had been shown effecting a cure for a patient who had suffered severe back pain for more than 20 years. The response to the programme was a clear indication

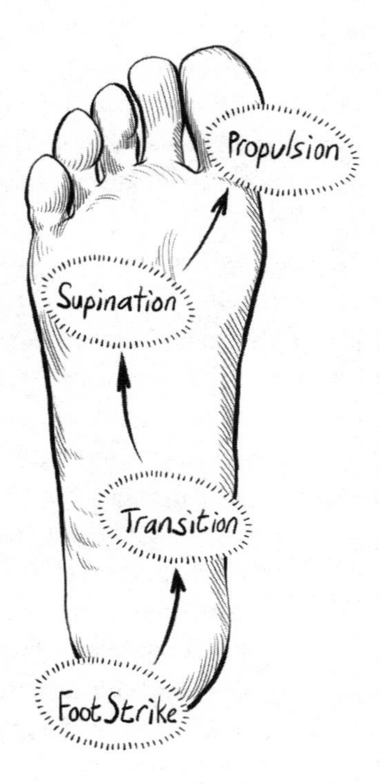

How the centre of gravity shifts over the dynamic ventral surface area of the foot during the gait cycle.

of just how many of us are in pain and in dire need of help.

Ward believes that most of us are unable to move correctly and that many biomechanical problems start in our feet. In the gait cycle, as we walk our centre of gravity shifts over different parts of our feet. We begin with supination at the heel (striking the ground at the outer rear edge), transition into pronation to make use of the arch mechanism, and then go into supination again at toe off (*see* diagram on page 211).

Ward also believes that many of our problems come from the fact that most of us "can't resupinate, extend our hips, or our spine." Resupination is the final phase of the gait cycle: toe-off when the foot moves outwards. The extension of our hips happens when our trailing leg is behind us during running (or walking). Extension of the spine is a movement that we might associate with traditionally "good" posture – standing up straight. Most of modern life is spent in flexion, with a curved spine and shoulders.

Many of us are under the impression that foot pronation (rolling inward towards the arch) is a bad thing, too. This rolling action of the foot, though, is there in the design and shape of the bones themselves. The base of the calcaneus (heel bone) is not flat and does not sit parallel to the ground. Instead, its lopsided shape encourages the foot to roll forward and move medially (toward the centre of our body) so it can use the arch mechanism to aid propulsion at toe off. In order to do this, the rear and the forefoot need to be able to move independently of one another, in the kind of motion made when wringing water from a cloth.

Ward says, "In the foot, all the bones have a three-dimensional axis. So each bone has to be able to transit in both directions, forward and backward, left and right and in rotation in order to be able to find centre, allowing the muscles some range of movement in all directions. That's centre. Without it, muscle tension increases; so does joint compression."

If something goes wrong (and it usually does) at least one joint will begin to move a little less, and another will compensate, with effects echoing up the closed chain until they find balance somewhere else. With every step taken, that pattern of change is locked into the joints, muscles and eventually the nervous system and the brain. This process happens alarmingly quickly.

Ward explains that if you bandage a couple of your fingers together "it only takes about two hours for your brain to start eliminating the idea that those digits could ever move independently. Imagine what being in a cast for six weeks does. By the time it comes off, you've forgotten what you can do with your elbow or your ankle, for example."

Is this why our feet are in such a mess – because they adapt to their environment, and especially shoes?

"If you ask me why our feet are a mess, it's because our bodies are a mess!" Flat-footedness is just the beginning of what might be going wrong further up the anatomical chain, so I ask him about the epidemic, particularly in developed nations such as the US, Europe and Australia, as well as some African countries (Kenya, in particular). Ward explains that we have to be wary of simple numerical assessments of movement health.

"There is this statistic which says 30 percent of our feet are normal, 70 percent abnormal. What that means is that the 30 percent are within normal biomechanical ranges, and those ranges seem to be 0–16 degrees. The problem is that if your left foot is 15 degrees and your right is 1 degree then you have a rotation in your pelvis, a knee that you can't extend and a spine that is rotating the wrong way; that's your set up. And it doesn't matter what the numbers are, they can still be 'normal', but the same problems related to gait patterns are going to emerge. So you've got to question what normal is; normal for the feet doesn't necessarily mean normal for the body."

The condition of our bodies is strongly environmental. The bodies we develop carry in their movement and their arrangement the mark of every trip, bash, fall, sports injury or other intense physical or harmful event, even going back to how you might have been pulled from the birth canal or delivered by C-section. Some osteopaths also believe that the posture of the mother can impact upon the biomechanics of their unborn child.

Our gait, in response to the wide range of life experiences, becomes as specific to us as our fingerprint. It is both like our DNA, and a beautiful expression of what happens to our DNA when it comes into contact with life and environment.

Foot posture is established by biomechanical strategies and intentions that emerge in response to life and this happens from the toes up. The items that feet mostly come into contact with in modern life are shoes. Ward is keen to avoid any idea that there is a perfect shoe. But he does believe that the ideal kind of shoe is "one that capitulates to the foot". He explains that the problem is that more than 95 percent of shoes dominate the foot instead. But what use are "barefoot" shoes to someone whose movement is already pathological?

"The barefoot shoe [a light, flexible shoe with no support of any kind] isn't changing the way people walk, it's allowing them to walk in the way that their mechanics are set up to do. And if their mechanics are shut down, locked, pronated, with limited range of movement, then what is the real restriction? Is it the shoe that is the problem, or the mechanics? The barefoot shoe might allow the right kinds of movement, but those mechanical strategies have to be already in place for it to happen. A really good tailored shoe (an old-school tailored shoe), for me is fantastic to walk in because my feet have got all the movement potential that they need, or they don't influence my gait too much. But when you go to a sports shop, most of the trainers sold there are full of gimmicks, whether it's shock-absorption that you don't need or promoting movement of the foot (which it can't do anyway), and most of them will create odd angles for the foot."

Ward feels that "the shoe is neither the problem nor the solution". And while I think that once the biomechanics are in place he is right, I also have a strong feeling that in the earlier phases of development, the shoe is part of an orchestra of problems that modern life presents us with. So what is the goal, or the solution?

"Your goal is to get the mechanics of your feet working optimally: to get the bones, joints and muscles doing what they do best; opening, closing, lengthening and shortening in order to help you move from a mobile adaptor (pronated foot) to that rigid lever (supinated) position as you propel yourself forward. There are no easy answers, but what I'm interested in is getting back to the raw basics of what any given structure is meant to do."

Strengthening feet, doing towel-scrunching exercises or heel drops, are probably not going to help the foot function properly. Many people's big toes have lost the ability to extend fully so they are unable to complete the final phase of the gait cycle: toe off. Strengthening this structure does nothing to return the range of movement that is needed to allow this function to take place. It is movement that we need, not strength and not stretches, but the ability to flow through a range of postures to completion so that everything in the chain can contribute to those simplest of things: propulsion, locomotion, movement.

The body's strategy to deal with a problem is to heal only to the level at which the part of the body works satisfactorily, rather than the level at which it works most efficiently and this creates an imbalance; that imbalance must be compensated for elsewhere; and what the body needs, it finds somewhere along the chain. The outcome of the body's healing strategy will not always result in pain. When it does, the source may be found in other sites of the body, away from the one that may be experiencing discomfort. It makes sense that for a body to be able to stand upright, everything has to be connected in a chain; if it isn't connected, we wouldn't be able to stand up.

When Ward said, "*Everything* needs to move" he might have been talking about bones, tendons and ligaments, and the complex negotiations they undergo between support and power, but never was a truer statement uttered. Everything needs to move. The idea sits at the very base of our long nominative classification as a species (*Animalia*). We only have brains because we move. Our bodies are wired up to reward movement in all sorts of ways, from delivering new brain cells, improving neurotransmitter ecology in the brain and adding life.

For millions of years, we roamed the savannah. For thousands of years we walked and farmed the land. Yet, what we have done for only a matter of decades in our sedentary revolution is what seems normal to us. The debilitating reduction in the strength and power of the human frame that is the result of office work is introducing pathology and morbidity at rates we have never seen before and could not have predicted. As a result, we use our feet probably less than we ever have done in the long history of our species, and "support" them on the few occasions when we do use them.

Moreover, the environment that our feet meet when we do use them is homogeneous in the extreme. The fact that everything is covered in concrete and tarmac means that we live in an artificially flat world, made all the less challenging by our wearing shoes. The effect is that every foot strike that we make is practically identical, and the muscles in our feet are rarely stretched or worked as they are when barefoot on "real" terrain.

Today, we are obsessed with our 10,000 steps (all of which are exactly the same) and our Air Maxes, unaware that we are barely doing what our bodies really need. If movement were a diet in modern life, we would be starving.

Once our feet start to go, the domino effect begins. It gets a little harder to stand up straight, the pelvis starts to tilt, lordosis begins to twang in the lower back, which makes it harder to exercise; we become more likely to develop chronic back pain, and behind all of these are an angry mob of pathologies who lie in wait for us to open the gates and let them in.

Office life may seem safer on first glance, but the risks we take when we climb into our first adjustable chair, switch on our computer and check our phone is plugged in, seem on reflection to be at least comparable to those taken by factory workers in the 19th century.

CHAPTER 8

THE WAIST LAND

The most poetical thing in the world is not being sick.

G K Chesterton, *The Man Who Was Thursday*

There are few people in the West unaware of the most common causes of death: cardiovascular diseases, neurodegenerative and metabolic disorders or cancers. Despite modern life's best intentions, we are living longer, which is one of a number of reasons why neurodegenerative diseases in particular are becoming more common: they are mostly expressed in old age. Alzheimer's and dementia are now the leading cause of death for anyone over the age of 85.

We are all expecting to live longer, aspiring to be like the folks in Japan, Hong Kong, Macau or Monaco, where if you're male you can expect to sail on through to your mid to late 80s. In the Western world we have been fairly competitive over life expectancy. These numbers can be misleading and need some adjustment for the kinds of palliative care available to those who can afford it in these countries.

In the UK, though, life expectancy has recently stalled. The last century added more than three decades to our lifespan. Life expectancy had been climbing so fast that it was anticipated that with each passing decade women could expect to add another two years and men another three. But in the last seven years it has stumbled. According to the Office for National Statistics, in 2010 life expectancies were estimated at 82.6 for women and 78.7 for men. By 2015, these figures should have been 83.6 for women and 80.2 for men. But both fell short; the actual figures were 83.1 for women and 79.6 for men.^

However, 2010 was the year in which National Health Service and social-care budgets were slashed as part of the new Conservative government's strategy for fiscal recovery after the global financial crash of 2008. This, coupled with the fact that if you are male and born somewhere in Africa – Angola, say – you will find you are only expected to live to 37 (women just, and I mean just, make it to 40). Where there's money, there's invariably greater longevity. As I learned from meeting Gianni Pes in Sardinia (*see* page 113), our genes are not terribly powerful determinants of our longevity, but it looks as though money certainly is.

Here are two maps of England that appear to represent nearly identical data. But what connects them is alarming. One is an economic map; the other a health one.

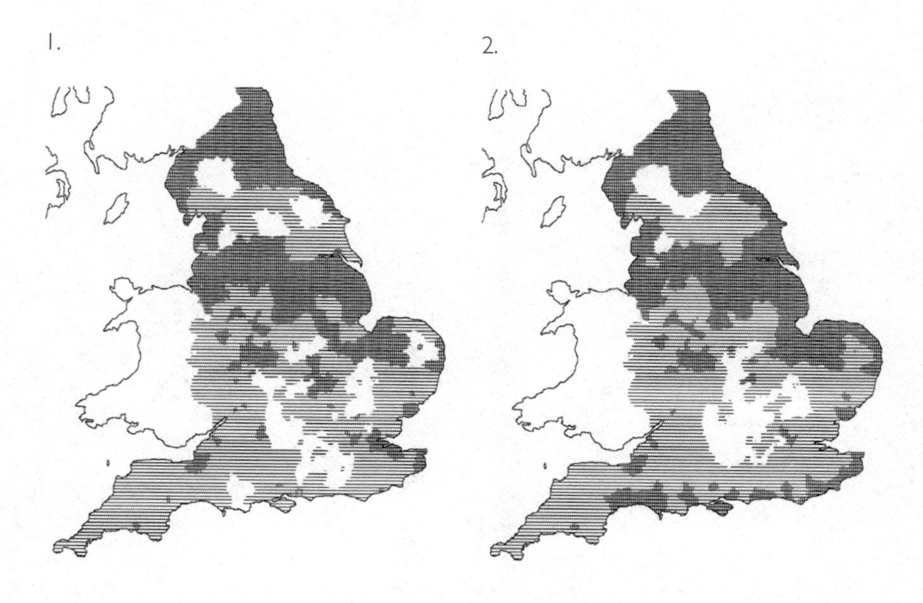

Map 1 indicates the risk of poverty and economic precariousness in different geographical locations. **Map 2** is based on a different data set and shows the risk of death from pulmonary obstructive heart disease in these areas. The darker the shaded area, the greater the level of risk.

The similarity between risk of poverty and early death is so similar as to look like a printing error.

Regions by wealth, the darkest areas indicate greatest poverty.

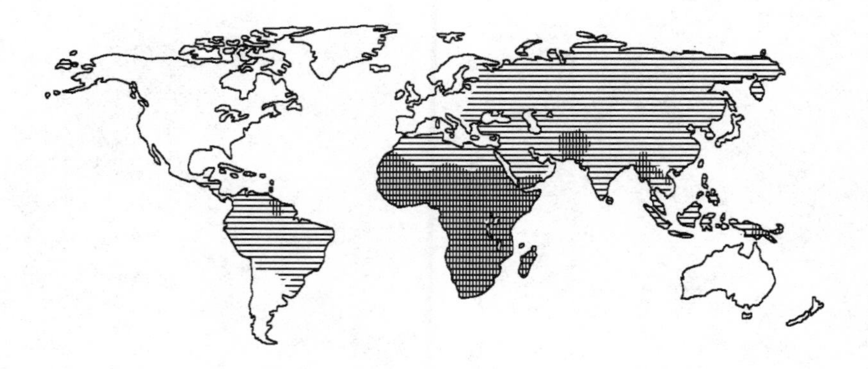

Regions by life expectancy, darkest areas indicate shortest lifespan.

You don't need to be Karl Marx to draw the conclusion that the material conditions of life affect consciousness; in this example, they seem to eradicate it permanently.

On a world map of healthy life expectancy a similar pattern emerges in which the relative health of a country aligns with some accuracy with the nation's income.

There will be an upper limit on how old it is possible for a human to be, but we are not there yet. In the meantime, as the science improves, as medicine continues to work miracles with palliative care strategies, life expectancy will take care of itself, and in the right soil, will continue to grow.

The much more pressing problem in the meantime is morbidity. The idea of getting old, to many, is utterly depressing. The thought of being prescribed medication that your life depends on, of taking pills just because old people need them, of being bundled into a mobility scooter, chucked out of your house and it being sold so you can pay your care bill, becoming so enfeebled that you can no longer leave your room, bathe, or go to the toilet by yourself – these are not inviting prospects.

But this is not everyone's future – and it is a future over which we have some control. Life expectancy is more than a little connected to financial status, although many of the elements that contribute to a good later life are associated with lower income, but not inevitably linked to it. Perhaps the principal one is diet.

Calories are cheap, organic produce is not – and is it the case that even the best organic food is all that nutritious?

Since the mid-20th century our waistlines have begun spiralling out of control. We are getting fatter because of an environment of calorific profusion mixed with an astonishing drop in calorific expenditure, particularly related to work, but also to modes of leisure that increasingly require us to sit still while we enjoy them.

For most people, about 40 percent of their weight (above or below normal) is genetically determined. But for some people, the percentage is much higher. Those genes work on the brain circuits and in hormones, such as leptin, that maintain appetite. This means that we do not experience satiation consistently across a population. And some unlucky people do not experience it at all. Others cannot burn fat. We all eat, but we don't all gain weight, and for some it is impossible to lose. We know from touching upon epigenetics (*see* page 99) that the 40 percent figure in terms of genetic-determinism is not huge, but it is substantial, especially when we consider the hostile energy environment that those genes meet in the Anthropocene.

Welcome to the Waist Land.

WHAT OUR DIET TELLS US ABOUT EVOLUTION

In 1880 British zoologist Edwin Ray Lankester was following in Darwin's footsteps and was interested in the destiny of species. His particular expertise was the history of invertebrates. His book, entitled *Degeneration: a Chapter on Darwinism*^ explored the different ways in which evolution might play out within any particular species. He was the first scientist to use the term that would go on to become incredibly popular within social Darwinism.

Lankester saw that species may:

1. **Evolve to a point of stasis, like molluscs, starfish, dragonflies...** "the process of natural selection and survival of the fittest has invariably acted so as either to improve and elaborate the structure of all the organisms subject to it, or else has left them unchanged, exactly fitted to their conditions, maintained as it were in a state of balance."

2. **Continue toward a state of ever-increasing elaboration (any user of iTunes will know exactly what this means: computer software is renowned for becoming ever more complex; hence the term "bloatware").** "Elaboration is a gradual change of structure in which the organism becomes adapted to more and more varied and complex conditions of existence."

3. **Finally, species may commence a process of degeneration:** "Degeneration may be defined as a gradual change of the structure in which the organism becomes adapted to less varied and less complex conditions of life."

Lankester provides several examples of molluscs, barnacles and crustacea to illustrate his points, but it is only that last definition that we are interested in. The idea that a species can degenerate when its food is too easy to come by is a troubling one. Surely that would mean that a particular species was at the top of its food chain and had evolved bodies expert in acquiring sustenance.

The smilodon was a sabre-tooth cat in the Pleistocene. Its teeth could be nearly 30cm (1ft) long. It was fast, agile; its jaws could open to about 120 degrees. Their structures were compact and robust – not leopard fast, but very strong. And they disappeared at the beginning of the Agricultural Revolution.

If Lankester could be pressed to speculate on their sudden disappearance, he would likely suggest a process of degeneration. This species was a hunter, clearly at the top of its game. Fast, tough, with sabres for teeth, the smilodon had clearly won the evolutionary draw. The smilodons' supreme adaptation for their environment probably made food sources a little too easy to come by. As the Pleistocene gave way to the Holocene, this species was not used to travelling long distances and was unpractised at stalking many different kinds of prey. Once their environment changed and disrupted their food sources, their elaborate form did not adjust well to doing things a little differently. Their degeneration almost certainly led to their extinction.

The question stands: *are we the sabre tooth*?

Our levels of inactivity have profoundly complex causes, but our environment has changed substantially. Like them, we are also standing on the cusp of a new geological epoch as we watch the sun set on the Holocene and await the heralding of the Anthropocene. Our bodies are too complex for what we require of them. We do not need so many joints and so much mobile potential; we need simpler bodies to perform the work we are currently doing, but our DNA does not know this.

Of the 7.3 billion people on the planet, more than 90 percent of us find food easy to come by. Many of us even have it delivered to our door. We find food with ease; we live near our places of work (mine is in the next room) – and if we don't, we climb into a car and drive ourselves there. Except for the fact that our teeth are getting smaller, I would be inclined to say that *we are* rather like the smilodon.

Animals in captivity that do not need to search for food are invariably heavier than their wild cousins for this reason. The signs are already here. In research recently published by Public Health England, the authors found that about one in two women and a third of men in England are damaging their health through lack of physical activity. More than one in four women and one in five men are classified as "inactive" because they do less than 30 minutes activity a week. In some local communities only one in ten adults are active enough to stay healthy and there are significant inequalities between different demographic groups (gender, race, disability and age). Physical inactivity remains one of the top ten causes of disease and disability in England and is responsible for one in six deaths in the UK. This costs the UK an estimated £7.4 billion a year.^

The report makes difficult reading, but there is a lot that can be done with such information other than scanning it with a sinking feeling. While there is no going back – we can't don a loincloth and head to the local park with a spear in the hope of bagging a squirrel – but what we can do is make small incremental changes to our lives, so that we are doing something that at least resembles what we might have been doing in the world that our bodies were expecting to find when cells packed with DNA began dividing to make new life.

All these factors are contributing to an environment in which opportunities for calorific expenditure are becoming so rare that only those who can afford it have both the time and money to burn energy and avoid some of the big killers, such as type 2 diabetes. As our waistlines grow, the disease and its complications are spreading fast. Its related costs are spiralling beyond our control. There is a historical term for the point at which a society in conflict turns all its economic production toward it: it's called Total War. Everyone and everything becomes subsumed into the war effort. With the numbers as they are, and with the rise in type 2 diabetes and obesity unchecked, the cost of treating these diseases will in a few decades overwhelm even some of the world's strongest economies, leaving those governments to make some ruthless decisions to deal with the new culture of "total obesity".

As things stand, obesity and its related diseases are estimated to be likely to cost the global economy $1.2 trillion each year by 2025.^ These World Obesity Federation figures also estimate that more than a third of the world's population will also be overweight or obese (2.7 billion of us). The figures vary from country to country, with the US leading the pack at more than half a trillion dollars. These are the sorts of figures that could bankrupt healthcare services.

Some have questioned whether obesity is a disease, but the numbers certainly suggest it is. According to the US National Center for Health Statistics, a chronic disease is one lasting three months or more, cannot be prevented by vaccines or cured by medications, nor does it just disappear^. In addition, a disease can be considered as chronic if it persists for a long time or is constantly recurring. Thus childhood obesity is not simply a disease; instead it is a chronic disease.

There is disagreement, though. Some physicians don't buy this. They don't like the idea that so many of us might be diseased, or that we are irresponsible, and state that it is a simple question of mathematics – eating too many calories means you get fat.

Obesity is a complex disease with many causes, none of which are "lack of resolve". Such perceptions belong in the mid-20th century. Instead, causes are now known to range from genetic, prenatal, biological, and psychological influences (over which the individual has only, at best, limited control). There are also more obvious social causes like early developmental influences. Money is an important factor, as is time, and the environment plays an important role in so many ways (by dictating habits and the availability of certain foods). For these reasons, barracking "willpower" from the sidelines is not a helpful response to what is fast becoming a global health crisis. Shaking one's head in disapproval leaves the world unchanged with all its problems firmly intact.

Obesity is increasingly linked with other kinds of environmental factors – microbiota, trillions and trillions of organisms in our gut that train our immune systems in early life (more on this on the following page), are being shown to have strong links to obesity. Research is demonstrating that the environment food meets in our gut changes the ways in which it is processed, meaning that certain blends of microbiota might be more effective at absorbing unwanted energy from food. Research published in *Science* in 2013 certainly suggested just this.^ In this trial, the gut flora from genetically identical twins was harvested (their poo was scooped). Though the pairs of twins were genetically identical, one was fat and one was thin. The microbiota from these two kinds of groups of twins was then transferred to a group of germ-free mice (bred with no micro-organisms in any part of their gut). The germ-free mice were then put on identical diets, with the same calorific intake and expenditure, but they did not stay the same weight. The germ-free mice that received gut flora from the fat twin put on more weight and accumulated more body fat than mice given the gut bacteria from the lean twin. In this case, obesity seems to be transmissible.

At the very least, the findings suggest that the interactions between body mass, diet and gut flora are more complex than we are currently able to show. The idea that something as simple as calorific intake is to blame for the obesity epidemic is beginning to look suspect and it might be that diet is more broadly to blame. Therefore it seems that the key is not just the foods we eat, but the fact that those foods have an impact on the way we digest and process all other food in our diet. Just as with neurotransmitters and depression, this looks like a sensitive system that does not respond well to new external stimuli.

Metabolism is not just about the ease and speed at which energy is used in the body, but also the way foods are processed by the digestive system. If you have a highly populated gut with a wide variety of flora you are less likely to retain weight (which is why there is no evidence of hunter-gatherer obesity). This is why recent studies are linking weight loss with the quality of the food rather than with calorific intake versus expenditure. A Stanford-based study in 2018 found just this. People who cut back on processed foods (especially sugars and refined grains) lost significant amounts of weight without fretting about portion sizes or calorie counting.^ The subjects in the trial also had their DNA analyzed to look at variants that operate on the metabolism of fats and sugars. Despite some variety among the subjects, their genes did not change the ways that they responded to their "healthier" diet.

The specific and individual nature of our gut ecology also goes some way to explaining why it is that on similar diets only some people go on to develop type 2 diabetes. Again, this is not just about genes, but about *two* environments – one outside, and the other a cosmos within – populated with more microbes than there are stars in the universe.

In future, there will no doubt be probiotic interventions; food-based therapies that will treat or prevent obesity. But you can't just go to the supermarket and grab a probiotic drink off the shelf and expect it to be efficacious. Mass-produced probiotics are about as effective as a mass-produced wig. To work effectively they have to be modelled, tailored and matched specifically to the individual. The same goes for faecal transplants. If you're a highly allergic individual, clearing some of your large intestine and replacing those bacteria with those of a non-allergic person seems like a reasonable solution.

A trial of gut microbiota transfer in 2016 explains that "it is possible to reproduce aspects of depressed behaviour and physiology via a gut microbiota transfer. This suggests that the gut microbiota could play a causal role in the complex mechanisms underlying the development of depression."^

As the research progresses it will no doubt reveal imbalances specifically associated with depression and other psychiatric disorders. Does the future lie in faecal transplantation? Whose faeces are you going to get? A non-allergic person's? Great! A thin person's? That would be good, too. A happy person's? I think that crosses a line. The whole procedure seems to be mucking about in an ecosystem it doesn't understand, and anyone who tells you they do, doesn't understand it either.

With the environment outside changing so much, it should come as no surprise that the unbelievably complex environment in our guts is also shifting. These delicate ecologies seem to lean on one another and when one of them is disturbed, its effects are felt more deeply than we could possibly have imagined.

"LET THEM EAT CAKE"

To the best of my knowledge, it was a 2012 article in *The Lancet*^ that first connected the mortality risks of sedentariness with smoking. Its authors explained, "Physical inactivity burdens society through the hidden and growing cost of medical care and loss of productivity. Getting the public to exercise is a public health priority because inactive people are contributing to a mortality burden as large as tobacco smoking." The similarities do not end there, they went on to explain, "Smoking and physical inactivity are the two major risk factors for non-communicable diseases around the globe. Of the 36 million deaths each year from non-communicable diseases, physical inactivity and smoking each contribute about 5 million."

Here is a ranking of causes for "disability-adjusted life years", meaning years of life lost plus years lost due to disability published in *The Lancet* in 2015.^

CAUSES OF DISABILITY-ADJUSTED LIFE YEARS

1. Dietary risks
2. Smoking
3. High BMI
4. High blood pressure
5. Alcohol and drug use
6. High blood sugar
7. High cholesterol
8. Kidney disease
9. Low physical activity
10. Occupational risks

In this chart, with the exception of 2, 5 and 10, disability-adjusted life years seem almost entirely governed by diet and movement. And if we had to choose between them, the greatest influencing factor would be movement. There is much intersectionality in the list: high BMI is associated with low physical activity, high blood pressure and high blood sugar (type 2 diabetes); and diabetes and high blood pressure in turn are both associated with kidney disease. These are only associations, though, not causes. We all recognize these causes of disability-adjusted life years. If we don't represent a few of the risks ourselves, then we know family members and loved ones who are affected by them.

If this were a list for *Homo ergaster*, *habilis* or even pre-agricultural *Homo sapiens* it would be completely different. Dietary risks would still be number one (because of paucity and lack of variety of food), but many of the others would not figure at all because they are Anthropocene diseases.

This cannot be stated too heavily: while these diseases have a genetic component, they are overwhelmingly exacerbated by lifestyle choices and the spectre of type 2 diabetes lurks behind many of them.

Like the other risks in this list there is a genetic component to type 2 diabetes that *Homo ergaster*, *habilis* and our other genetic elder siblings would no doubt have shared. Diabetes is a disease in which the body's ability to produce a hormone called insulin is either impaired or is not able to respond effectively. All animals produce insulin (and even some mushrooms manage to do it). When you eat, your pancreas releases insulin which sends a message to your energy stores (muscles, liver and blood) to absorb the glucose cell by cell. Insulin is the key which unlocks the body's cells, allowing the transfer to take place.

Type 1 diabetes is much rarer than type 2. Type 1 is an autoimmune disease which affects the production of insulin in the pancreas (the immune system destroys the beta cells in the pancreas that make the hormone). It generally develops during childhood. Rates of type 1 diabetes are not as high as type 2 (which make up about 90 percent of cases). The causes of type 1 are not understood and experts believe that there are genetic and environmental factors (if one identical twin develops T1, there is only a 30–50 percent chance that the other will). At the moment, there is no known method for preventing the onset of T1.

Nonetheless, the figures do suggest environmental links. A 2014 study published in *JAMA* showed that from 2001 to 2009 there was a 21 percent rise in cases of T1.^ Given the prevalence of other sudden rises in diseases in the Anthropocene, an environmental link seems more than likely, especially in the light of research that is beginning to focus specifically on that emerging field of the Old Friends hypothesis (*see* page 242).

Another study published in *JAMA Paediatrics* in 2014 concluded that although more research is needed, "early exposure to an indoor dog may protect a child from preclinical type 1 diabetes."^

Type 2 diabetes is a great monster of a problem. Unlike T1, it is clearly associated with obesity. With weight gain, the cells in our bodies develop an increasing resistance to the effect of insulin, so for the hormone to work we have to produce more and more of it.

Glucose has to pass from the blood and into our cells so that it can be used as energy, and insulin assists its movement. But when we carry more weight the increase in fat, particularly around the abdomen, means that our cells get a fatty sheen, like waterproofing a jacket by waxing it. This makes it harder for the insulin to penetrate the cells, and consequently the cells cannot absorb blood glucose which builds up in ever-higher concentrations, circulating around the body trying to find a cell that isn't waxed over with fat. This is called insulin resistance. But it is not a level playing field. Just being overweight does not mean you will become diabetic.

Insulin resistance increases a little with age, but abdominal fat has the strongest associations with the disease. It is not about whether you eat a carb- or protein-heavy diet; it is about energy input and output. Excess calories (whether from sugary foods, vegetables or meat) will eventually lead to higher concentrations of fatty acids in the blood, which increase insulin resistance.

This is why the processing of food in our culture is such a problem. Not only has it changed the shape of our faces (causing malocclusions where none might have existed), and not only are micronutrients being depleted and good bacteria being killed, but sugar is often added to it.

This is not in itself a problem, but it quietly increases the energy impact of food already streamlined to deliver a maximum calorific fatty sugar bomb. In this food, artificial additives are so weird and inexplicable that I don't think anyone without a PhD in chemistry could tell you what half of the ingredients are in a ready meal. It's not that the refined carbohydrates in processed foods can be broken down a little too easily by the body (leading to spikes in blood sugar and insulin). It's not that they are lower in nutrients, vitamins and minerals than whole foods. It's not that processed food has much less fibre. All of these are good reasons to moderate the quantity of processed foods we eat, but the sum effect of such characteristics is that our diet seems to consist of a whole lot more, but provides much less of what we need. It's both fuller and emptier. A processed meal gives us less of a nutritional bang but a whole lot more calories and salt. Our bodies also process it quicker (because some of the processing has already been done elsewhere), which means we feel hungry again sooner.

Other processed foods cast rather long shadows over our health, too. The World Health Organization provides shocking advice on processed meats. Processed meats include ham, bacon, sausages, hot dogs, corned beef, chorizo or any meat that has been transformed through curing, fermentation, smoking, salting, preserving or other processes that enhance flavour. They are ranked as category one carcinogens, along with cigarettes, asbestos, arsenic and plutonium. The health risks associated with processed meat range from increasing the chances of developing type 2 diabetes (one portion a day increases the likelihood by about 50 percent) but also of developing cancer.^ A number of studies attest to the fact that eating only a 50g (1.8oz) portion of processed meat a day (about a tablespoon) increases the risk of colorectal cancer by about 18 percent.

Dr Margaret Chan, Director-General of the World Health Organization, has called the obesity and type 2 diabetes epidemic a "slow-motion disaster". Advertising, distorted science, agricultural and trade subsidies, government lobbyists and international trade policies (the undercurrent of all of these is money), have conspired to create what is effectively an obesogenic environment in which calories are cheap and burning them is expensive. It is not just that gym memberships cost money; walking to work takes time and preparing food does, too. Time is more costly than most things that we otherwise find the money for.

The pharma companies are onto this problem, though. A new pill will soon be up for FDA (Food and Drug Administration) approval in the US^. The drug, GW501516 (nicknamed 516) is already banned by the World Anti-Doping Agency and is available on the black market. It is a synthetic replacement for the kind of chemical activity stimulated by exercise. The drug makes the body burn more fat, and makes it easier to metabolize sugar, facilitating the loss of more body fat; it works on endurance, too, so exercise can be done for longer. But its main function is that it tricks the body into thinking it has done exercise when it hasn't.

The clinical applications for the drug mean that it might be highly beneficial for people not able to exercise: the chronically obese, sufferers of severe muscular dystrophy or quadriplegics, for example. This is a drug, though, that has been under development for at least 11 years, and it seems ripe for abuse. Exercise is already symptomatic of a lifestyle that has gone awry and become too sedentary, and this drug aims to streamline exercise so that less of it can be done to supposedly greater effect. The drug also denies the consumer most of the psychological benefits of exercise, and because it encourages lower levels of activity, bone density-making osteoclasts are less stimulated, which could lead to even more brittle bones.

One of the other effects of the drug is that the mice the drug has been tested on were better able to control their blood sugar levels. But do we need a pill for this? Type 2 diabetes is more preventable than many of the diseases we have looked at in this book. Decreasing calorific intake will help, and insulin resistance can be lowered by losing as little as 2.2kg (5lb), or going for a 10–15 minute walk every day (a few months of this habit would burn off well over that much weight). This may not sound like much, but a little goes a long way.

Type 2 diabetes is a little like short-sightedness; if left unchecked it will rage rampant. A recent trial run by Imperial College London and Harvard's School of Public Health looked at data from 4.4 million adults from most countries and they estimated the annual cost of type 2 diabetes to be about $825 billion per year globally (China $170 billion, the US $105 billion, India $73 billion and Japan $37 billion). They found that rates of type 2 diabetes had nearly quadrupled since 1980. In 2014 the disease affected 422 million people.^

The number of people suffering from type 2 diabetes is anticipated to climb as high as 800 million in the next ten years. The Anthropocene body exhibits numerous signals of stress in environments of abundance. For more than 2 million years our species has fought predators and starvation to stay alive; today we mostly die because we eat too much.

But what if the Anthropocene has an additional and pervasive role to play in all this? What if it was interfering with the growth of food itself, so that everything produced more sugar than it had in the past? What if the world we have made was changing organic food into a new kind of junk food?

"LET THEM EAT DIRT"

Question: when is a carrot not a carrot? Imagine two carrots. Both are grown in the same kind of soil, in the same field, using the same kind of organic matter as fertilizer, in the same climate. When they are pulled from the earth, the only thing to differentiate them is that one is from the present while the other was harvested two centuries ago – same size, same weight, probably the same colour, same length of time in the soil. But when they first saw daylight, were peeled and later eaten, which had the most nutrients and which the most calories?

For decades, researchers have noticed that our food is becoming less nutritious. If you think about it, this seems to make sense. We use more intensive farming methods, the crops that are grown are done so for yield, not necessarily for nutrient content. This itself is not news. But what if this is only part of the story?

What is happening to our food in the Anthropocene touches on questions of global health in ways simple enough for a schoolchild to understand, but the effect is nonetheless far-reaching.

The answer lies in a simple and magical equation that you probably learned at school:

$$6CO_2 + 6H_2O = C_6H_{12}O_6 + 6O_2$$

This is the equation for photosynthesis: the means by which plants use sunlight to turn water and CO_2 into glucose. The equation tells us that carbon dioxide plus water exposed to sunlight (and with the help of a few plant enzymes) makes sugar (for the plant) and oxygen for us (a waste product for the plant). Because of their magical and transformative abilities, plants don't need to work, walk or hunt; they are self-sustaining. They make their own food from water, CO_2, nutrients from the soil and sunlight.

The atmosphere of the Earth in its early days consisted of carbon dioxide and water vapour. About 3.2 billion years ago, the evolutionary process struck on this self-sustaining method for plants and over about 800 million years the atmosphere became oxygen-rich and the Earth gradually took on its recognizable blue and green appearance. This is called the "oxygen revolution" (yet another one).

Before this period, single-cell life predominated for several hundred millions of years. One of the key things to realize is that in photosynthesis, oxygen is a waste product. Oxygen is pumped out of the exhaust of the photosynthetic process.

Oxygen is a volatile gas, and was poisonous to the single-cell life that predominated on the planet. We live off a gas that for quite a span of the Earth's history was an environmental pollutant. CO_2, on the other hand, is a key plant nutrient, and increased levels result in greater plant growth. This is why there have been series of experiments on growing food in CO_2-enriched environments. CO_2 is used a bit like a fertilizer to encourage growth in crops – and it works, or seems to.

Fast-forward a few billion years and the world has changed and remained the same. It has been through a number of geological cycles (such as ice ages), but the components of its atmosphere on a larger scale are similar: the air is still oxygen-rich and supports multi-cellular life.

Ecosystems are delicate things, though, and the balance has begun to shift.

Since the Industrial Revolution, the levels of CO_2 in the atmosphere have risen. When the dinosaurs roamed the planet, levels were as high as 1,000 parts per million (ppm). But we evolved in an atmosphere that had much lower concentrations, about 200–300ppm. There is always natural fluctuation in the Earth's levels, but it tends to be moderate, about 20–40ppm.

Today, the average is already over 400ppm, and the Intergovernmental Panel on Climate Change has estimated that by the end of the century, in various scenarios it will be between 550 and 900ppm, which is great for hubristic bearded old men who have sunk their life savings into a dinosaur theme park, but not so good for the rest of us.

Scientists estimate that CO_2 levels will climb as high as 550ppm by the year 2050. The literature on the effects of these levels of carbon dioxide on our climate if left unchecked is substantial. But it is also having an impact on the content of our food – a problem most of us know little to nothing about.

Carbon dioxide is a key component in photosynthesis, so what happens when there is more of it in the atmosphere? If bigger numbers are pumped into the equation, then bigger numbers come out. More CO_2 in, more plant matter out. It is for this reason that a Republican representative from Texas, Lamar Smith, in an article entitled "Don't Believe the Hysteria Over Carbon Dioxide" argued that "a higher concentration of carbon dioxide in our atmosphere would aid photosynthesis, which in turn contributes to increased plant growth. This correlates to a greater volume of food production and better-quality food. Studies indicate that crops would utilize water more efficiently, requiring less water."^

Nearly all of this is correct, except for the bits which aren't, where he goes wrong before he goes a-whole-damn-world of wrong. The bits of his statement that are true are "contributes to increased plant growth" and, "correlates to a greater volume of food production". These parts are correct, partially.

I recently heard of a study compiled by a mathematician, Irakli Loladze, and published in *eLife* in 2014.^ He writes about ionomes. An ionome constitutes the total elemental composition of any organism, so the paper was assessing the impact of raised CO_2 levels in the Earth's atmosphere on the elemental makeup of plants.

While studying for a PhD at Arizona State University, Loladze modelled a system in which the quality of photosynthetic food affected the animals that ate it. After moving to Princeton University, he raised this problem of accelerated photosynthesis and changing food quality but couldn't generate much interest. A researcher in the Department of Geosciences told him that if levels of a nutrient as important as iron dropped it didn't really matter "because people can always eat a little bit of dirt". Lots of animals do this, elephants and even goats are committed rock lickers, but humans are likely to be a little more sceptical and refined in their tastes.

In 2002, Loladze published a paper on ecological stoichiometry: the balance and the ratio of chemical elements essential to all life forms. He asked: if plants draw CO_2 from the atmosphere but draw almost all other chemical elements from the soil does a problem arise because the rise in CO_2 in the atmosphere is not paired by a perfectly-matching equivalent enrichment in the soil?^ The elements in the soil do not change just because the atmosphere has a few more parts per million of CO_2. As extra CO_2 in the atmosphere pushes the accelerator pedal for the production of glucose in the plant, there is no equivalent acceleration of nutrient uptake. This suggests that there may be a global elemental imbalance that could have an impact on chemical elements such as "iron, iodine and zinc, which are already deficient in the diets of half of the human population". He was trying to ascertain whether CO_2 made plants and vegetables less nutritious and more calorific.

It's difficult to overestimate the importance of this question when 40 percent of *all the calories consumed by our species* are made up of rice and wheat. The answer could mean that a loaf of bread from the 18th century was more nutrient-dense than any loaf you might find anywhere on the planet today.

The 2002 article was theoretically fairly conclusive. Loladze was able to deduce that even though plants might have grown larger, yielding more calorific energy, "the concentration of nearly every element, on average, should decrease."

The way his work differed from that of his peers was that he linked changes in the quality of plants grown in an elevated CO_2 atmosphere to human nutrition. He attempted to draw from all the published data on CO_2 and essential chemical elements in plants: nitrogen, phosphorus, potassium, calcium, sulphur, manganese, iron, zinc, magnesium and copper. The problem Loladze immediately came up against when testing his hypothesis was the paucity of data.[1]

Over the next ten years, with no lab, and only a laptop to work on, he was

1 The data set was too small relative to all the "noise" (variability) to empirically confirm his logical derivation, or "thought-experiment" as he called it.

able to compile a data set comprising more than 15,000 data points, a big improvement over the data in the 2002 article, and to date this is the world's largest data set on the topic. Loladze had an empirical proof that rising CO_2 consistently lowers the quality of plants and published his results in 2014.^ In that study, Loladze noted: "Human activities profoundly alter the biogeochemical cycle not only of carbon but also of nitrogen, phosphorus and sulphur, which are central to all known life forms." It is plausible that other subtle global shifts in the physiology and functioning of organisms lurk amid highly noisy data.

Loladze explained that the "hidden shift" of his title was his umbrella term explaining that the drop in nutrients is "hiding" in all the data noise – it takes a very large sample size to detect this. But it is there; it is everywhere – on the four continents the data came from, in temperate and tropical latitudes, in crops and wild plants – and eventually it finds its way on to our plates. With increased photosynthesis, what we get is more carbohydrates and starches in the plants: "you sacrifice quality for quantity". We get extra growth in the plant, but "with the elements, pretty much everything drops. It was very clear to me what was happening. There are several mechanisms at play: I expected the minerals would drop and carbohydrates increase. But there is also another mechanism that takes place when CO_2 goes up. The plants take up CO_2 from the atmosphere through things called stomata on the underside of their leaves. If we think of them like little mouths, then when CO_2 is more prevalent in the atmosphere, plants close their stomata a little. They do so because when the stomata are open, the plant loses water because moisture in a leaf is higher than in the air. When the plant narrows the stomata, they lose less water, which means they *use* less water. They become more efficient; but, because they use less water, they also draw less water to their roots – and that means there are fewer nutrients in the soil flowing towards the roots."

Some of the other uncomfortable science that he is utilizing to assemble his case are studies such as one by Klementidis *et al* published in *Proceedings of the Royal Society Biological Sciences*.^ He tells me that studies like these are demonstrating that humans are not the only species gaining weight in the Anthropocene: "studies show that wild animals are getting fatter even though they have no access to human-produced food."

In Loladze's 2014 paper, he put together a final average for this upcoming nutrient famine "across all the 25 minerals, the mean change was −8 percent". How significant is this figure? "It's huge!"

"Even if it is only 5 percent, it's not one nutrient that drops, it's all of them. Suppose you live in a developed country where there is a 33–45 percent risk that you are calcium or magnesium deficient (which is true). Now, you get another 5 percent less of these nutrients from plant products and because minerals cannot be destroyed or created, it means there are fewer nutrients on your plate. You can't synthesize magnesium in your body. So either you suffer the consequences of the deficiency, or you increase your food intake. This kind of eating, called compensatory feeding, has been observed in animals. When you feed animals plant food grown at elevated CO_2 levels, they eat more. It's reasonable to assume that humans would also eat more. A mathematical model developed by Hill argues that if you eat 5 percent more calories at every meal, all other things being equal you will become obese within three years and morbidly obese within about ten."^

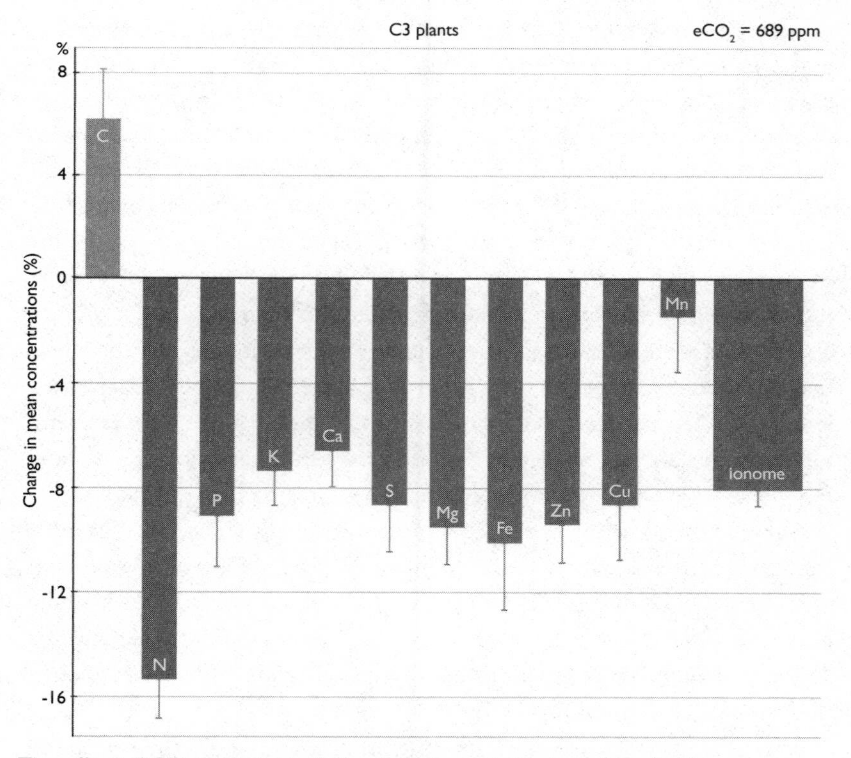

The effect of CO_2 on individual chemical elements in plants, from Irakli Loladaze, "Hidden shift of the ionome of plants exposed to elevated CO_2 depletes minerals at the base of human nutrition", in *eLife*, 2014.

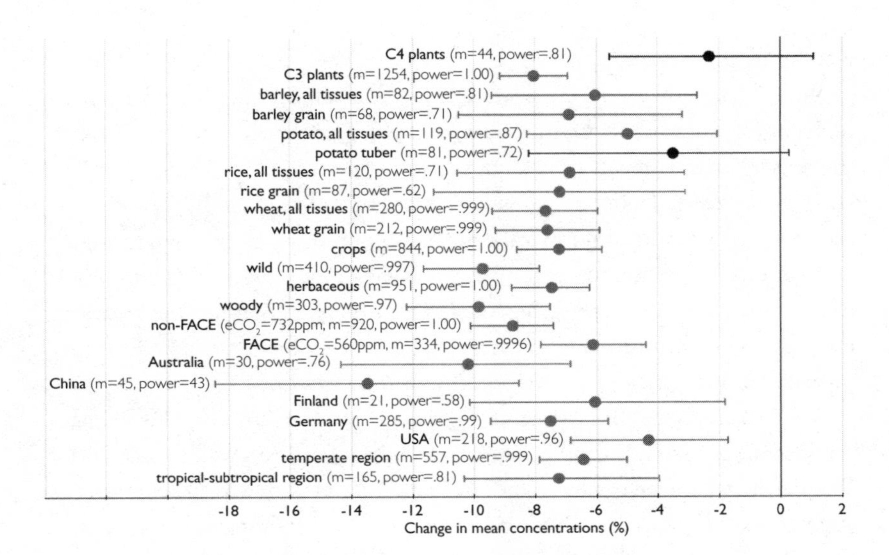

The systemic aspect of the CO_2 effect – change (%) in the mean concentration of minerals in plants grown in eCO_2 relative to those grown at ambient levels, in Irakli Loladaze, "Hidden shift of the ionome of plants exposed to elevated CO_2 depletes minerals at the base of human nutrition", in *eLife*, 2014.

The chasing of ever higher yields of crops is obviously well-funded (the Bill & Melinda Gates Foundation are among its sponsors), and while it will certainly yield more calories, it looks as though CO_2 is converting our plants and vegetables into junk food.

The implications for human health are as clear as they are shocking. I don't need to go into the sorts of havoc that can be caused by basic mineral deficiencies in the body. Loladze told me that there is not a single data point yet in existence that looks at what happens to lithium levels in plants grown in an elevated CO_2 atmosphere. Humans obtain lithium from grains, vegetables and drinking water. "Lithium is essential for human nutrition. It's important for mental health, it's needed for the delivery of vitamins like B12, and it's also needed for the creation of stem cells. We already know that lower concentrations of lithium in drinking water have been associated with higher rates of suicide. Very little data exists for other essential elements like selenium, chromium, iodine, molybdenum; the list goes on..."

Our ability to store fat for us to use during leaner periods evolved many years ago in an icy climate that froze and thawed unpredictably. During austere times, those few who were most able to turn surplus calories into fat were most likely to survive.

We did not evolve in an environment where our food was grown with such high concentrations of CO_2 in the atmosphere. During the next few decades, the pre-industrial levels of CO_2 in the atmosphere are set to double. It is still too early to tell what its global effect will be but the fact that it is already reducing the protein and mineral concentrations and ceding space to sugars and starches in an already carbohydrate-heavy diet is alarming. Surely it is not just me who wonders to what extent the obesity epidemic is being driven by a human appetite craving micronutrients that are being pushed out of the food chain by rising levels of CO_2 in the Anthropocene.

Marie Antoinette famously said of the masses that if they don't have bread, "let them eat cake". Eating cake has not worked out too well for us, and if Loladze is right, even organic vegetables are well on their way to becoming as sugar-filled as a *macaron*. Every bite of plant-based food for the rest of your life will be progressively less nutritious because of rising CO_2, and not only will it have more calories in it, but you will also need to eat more of it.

Now, to return to my original question: when is a carrot not a carrot?

...When it's an Anthropocene carrot. Its 18 chromosomes remain unchanged, but the way the carrot's DNA operates in this new environment has changed its body – does that sound like anyone else we know?

Our food environment probably plays the strongest role in driving the obesity epidemic and it has a number of causes. The most obvious is that our foods are, to a large extent, industrially produced and deliberately engineered to drive appetite by containing higher amounts of fat, salt, sugar and additives than they previously did. The relative cost of different kinds of food has meant that organic vegetables have increased in price, whereas these mass-produced foods are ubiquitous, cheap, and – if they are not already prepared – easy to cook.

This is our obesity epidemic, but the problem is that it is impossible to isolate a single driver among all the shifting causes. The fact that our environment changes the calorific density of our food, making it less nutritious by comparison, is yet another alarming variable to add to the complex algorithm of obesity.

By the year 2100, global population is estimated to rise to 11.2 billion. The Earth's atmosphere will suffer as CO_2 levels rise, but my, the carrots will be huge.

An Anthropocene carrot, spotted in Rome, 2018.

CHAPTER 9

DROWNING IN AIR, OR "PRACTISING HOW TO DIE" IN THE 20TH CENTURY

Pollution! All around
Sometimes up; sometimes down,
But always, around.

"Pollution" by Rik, from *The Young Ones*

... suddenly it resumed the attack ... I do not know why I should call it by its Greek name; for it is well enough described as "shortness of breath". Its attack is of very brief duration, like that of a squall at sea; it usually ends within an hour. Who indeed could breathe his last for long? ... anything else may be called illness; but this is a sort of continued "last gasp". Hence physicians call it "practising how to die". For some day the breath will succeed in doing what it has so often essayed.

Seneca (4BCE– 65CE) Letter LIV "On Asthma and Death"

One of the key drivers in atmospheric CO_2 levels is, of course, pollution from the use of cars. They are surely the symbol that dominates the 20th century. From E M Forster's early-adopting Wilcoxes puttering through empty country lanes in 1910 to the 35,750km (16,000 mile) Pan American highway which daily carries 420,000 cars between Canada and Peru, cars have become a part of everyday life and so have the smogs they produce by being an essential ingredient of pollution more widely.

In the early 1940s in Los Angeles, the smog got so bad that people believed it was some sort of chemical attack by the Japanese. In Pennsylvania in 1948, smog caused illness in hundreds of people and killed 20. In London, the creeping "red rust" had by 1952 become the Great Smog and it killed between 4,000 and 12,000 people, whilst 150,000 were hospitalized.

An international team of researchers (some from Texas A&M University) have recently been looking into the potency of the London fog and they discovered that sulphur dioxide (a product of burning lower-quality coal) was nudged into becoming sulphuric acid with the help of high levels of nitrogen dioxide in the atmosphere.^ The fog of 1952 remains the worst air pollution event in European history.

If this all sounds like the distant past, the same chemical corruptions are regularly cooked up in cities such as Beijing and Xi'an; though less acidic than the 1952 fog, the compounds formed are still potent and life-threatening.

It is easy to assume that the rise of asthma in such an environment is caused by pollution. But the evidence suggests that while it is an aggravator of allergic conditions such as asthma, it is unlikely to be the cause.

Until the 19th century, asthma was an exceedingly rare complaint. Vivaldi suffered with it, as did Beethoven; both would likely have been treated with bloodletting, herbal concoctions or been prescribed tobacco to smoke. Using our literary dipstick to look for its commonality in pre-industrial literature, a quick glance at Shakespeare reveals a very round zero. It is mentioned in Homer's *Iliad.* But because the early Greek term *asthmaino* meant "panting" it is hard to know if it is used to describe a symptom of breathlessness or for the disease itself.

In Ancient Egypt people might have tried to treat it with enemas, herbs, figs, grapes or the leaves of the belladonna plant. In Ancient China, people might have been given a tea which contained ephedrine (a key ingredient in a modern asthma drug). In India people might have tried inhaling cinnamon, which is a lovely thing to do, but probably not very helpful. Or they might have tried steam, turmeric, insect resins or any number of disobliging concoctions.

During the medieval period, some symptoms and triggers began to be recognized. Jean Baptiste van Helmont (1579–1644) a physician in Belgium, worked out that the cause of asthma originated in the bronchi; he called it an "epilepsy of the lungs".^ He also noted that inhaling dust, or, amazingly, fish, could bring on attacks. For these insights, and because they contradicted the Church's official line on health and the four humours, van Helmont was condemned to death (so had to recant his ideas). His work was never published in his lifetime. He retired from the Church and went on to lead a quieter and more restful life. Only on his deathbed did he give his books to his son and ask for them to be published.

A little later Bernardino Ramazzini (1633–1714), an Italian physician who became one of the first ever occupational health specialists, wrote about several patients who had come to him reporting irritation of the eyes. This led Ramazzini to conclude that there may be problems in their workplaces, which he investigated. He found that the workplaces had two major effects on the workers' "unnatural postures of the body", also noting the "harmful character of the materials that they handle, for these emit noxious vapours and very fine particles inimical to human beings and induce particular diseases".^

It is hard to underestimate the significance of a discovery as simple as this, but Ramazzini found that these porters, farmers, orators, athletes and runners, ragmen, grain sifters, stone cutters, millers, bakers, tanners, gilders, miners, tinsmiths, lime and gypsum workers, alchemists, potters, tobacconists, painters, copper smiths and printers were all at risk from asthma in their workplaces and from their working practices. Ramazzini was aware that in almost all cases, asthma was caused by sudden bursts of exertion (athletes in particular), but mostly by particles "sharp and acid" that adhere in the lungs. Throughout his writing he bemoaned the fact that although the workforce hated their jobs and their conditions they did nothing to improve them.

Ramazzini's book, *The Diseases of Workers* is widely considered to be the first work of occupational medicine, but for 150 years it was also the only one. Ramazzini was considered a joke by his profession, and it was not until about 1940 that his ideas became more widely adopted.

Despite these ancient and more recent histories of asthma and their symptoms and cures, its relative presence in 18th and 19th century culture is still on the introverted side.

Even as late as the 1950s, asthma still only affected less than one percent of the population. What has changed since then to make our environment turn on us as it has?

The number of asthmatics did not spike until the 1980s – and boy, was there a spike then. Over three decades, asthma numbers made a tenfold jump from 1 to 10 percent. But thanks to lots of environmental legislation, the air became cleaner during that period.

Genes seem a strong candidate for investigation. A review article published in *Immunological Reviews* in 2011 suggested some bewildering conclusions as to the genetic causes of common allergic responses. After combing through mounds of existing research, the researchers found that the many previous trials were unable to pinpoint genetic causes for these allergies. Instead, estimates varied: "between 35 and 95 percent for asthma; 33 and 91 percent for allergic rhinitis (hayfever); 71 and 84 percent for atopic dermatitis (eczema)." In each case, the ranges clearly signal that there is no genetic smoking gun for extreme allergic responses to our environment.^

Genetics, then, cannot account for the fact that 25–30 percent of Westernised Anthropocene humans suffer hay fever, asthma, eczema and food allergies (for example to nuts, milk, eggs or gluten) as well as strong allergies to animals with fine dander on their fur, including cats, dogs and horses. In hunter-gathering tribes such as the Hadza of north-central Tanzania, the rates of allergy are outstandingly low in the population. In many pre-hygiene African peoples, the idea of allergy in any form was more than a little unusual. There is no direct translation in Swahili (the lingua franca of East Africa) for "allergy".

In fact, the Anthropocene human is 450 times more likely to develop allergies than hunter-gatherers.

MICROBIOTA & THEIR ROLE IN DEVELOPING IMMUNE RESPONSE

Why are our bodies today rejecting environments that were previously essential for our wellbeing? Is it because we have become too clean?

The idea that our modern way of living is pathologically sterile was popularized by the Hygiene Hypothesis in the 1980s, which suggested that the lack of exposure to certain infectious agents, parasites and other microorganisms was causing us to develop allergic responses to harmless substances found in modern life. It's a "biome depletion" theory; the idea that we are daily scrubbed clean of bacteria and microbiota (the micro-organisms of a particular site or habitat) leaving our immune systems confused and malfunctioning, hypersensitive to the slightest stimulation and fighting battles that need not be fought.

When the research was initially published in the *British Medical Journal* in 1989, it noted that children from larger families were less susceptible to allergies than only children. It concluded that children from larger families were exposed to comparably more germs and infections via their siblings.^ It's a neat hypothesis, but once extrapolated, begins to look a little less persuasive.

Asthma and eczema were on the rise from the 1980s onward, but most of the major hygiene changes to our environment in Europe, Australia and the US had been established as long ago as the 1920s. Moreover, allergic asthma was on the increase in hygienic and unhygienic cities alike. Migrants into major cities in Europe and the US show fewer allergies than their natives – but their offspring don't fare so well and quickly show similar symptoms to their fellow city dwellers. It seems that city life does not agree with us, and especially not with our offspring.

Professor Graham Rook, an immunologist at University College London's Centre for Clinical Microbiology, believes that the problem is that our immune systems are ill-educated. Rook is the brain behind the Old Friends hypothesis, which is an alternative to the Hygiene Hypothesis and is fundamentally about three things: microbiota, the natural environment and old infections.

Microbiota are ambient species in the natural world that co-exist with and within humans. They are species that inhabit our gut, skin, lungs and blood, and they are also found in other animals.

The importance of the natural environment and microbial exposure has been proven in numerous studies. The most influential of these is "Innate Immunity and Asthma Risk in Amish and Hutterite Farm Children" published in 2016.^

The second was unknowingly set up by Stalin when he divided territory between the Finnish and Russian Karelia in 1939. A study published in 2015 explored how the allergy rates of two different sets of people – genetically similar but socio-economically distinct – responded to their different environments. In both studies, the groups that were less industrialized and relied on rural economies (the Amish and the Finnish Karelians) were protected by their environment against allergies and shaped a more appropriate immune response (which usually means no response at all).^ We know that allergies such as asthma and hay fever are caused by inflammation in soft tissues, which is an inappropriate response from the immune system to what should be harmless organic matter. Most discussion of the immune system concerns its activity, but 99.99 percent of its role is not reacting aggressively to things.

The Old Friends hypothesis suggests that the immune system has not been effectively socialized and is as aggressive as a frightened dog retaliating against the slightest stimuli. Our immune systems become hostile to what should be harmless substances like nuts or particles of dust or pollen. The immune system does not know they are innocuous because it has not been sufficiently educated at key moments of our development, right from birth. The womb was once believed to be a hyper-sterile environment, though it has since been discovered that components of the maternal gut bacteria already modulate the foetal immune system in utero. But our first high dose of friendly bacteria is delivered when we are born. As we travel down the birth canal and out through the vagina, we get covered from head to toe in friendly bacteria (mainly lactobacillus) as a first kind of immersive training for our immune systems. But not all of us are, as Lady Ga Ga famously put it, "born this way".

Technological interventions in the ways that women give birth have saved tens of thousands of lives. Caesarean sections are now more popular than ever. In a number of cases these are emergency procedures, but increasingly, they are elective. Thirty years ago in the UK, one in ten children was delivered by C-section; now the number is more than one in four, with more than a third of C-sections being elective. Those of us who are born by C-section miss out on that first mega-dose of friendly bacteria.

A Norwegian study of 1.75 million children in 2008 showed that babies born by C-section were 52 percent more likely to go on to develop asthma.^ (It is rumoured that it is now common practice among obstetricians to smear a little vaginal bacteria onto the mouth of the baby during such procedures. This is called vaginal "seeding".)

Rook also explains some of the important functions of human breast milk. "Breast milk was long thought to be sterile, but it is not. There are organisms in the milk and there are also polysaccharides that feed and encourage growth of certain gut microbiota." This substance, the perfect food for newborns, contains peculiar complex sugars that are indigestible to the infant.

The sheer ingenuity of the human gut is astonishing. Why is metabolic effort expended upon the making and delivery of a substance that contains useless ingredients? "It turns out that the polysaccharides are not for the infant at all. The organisms that, for developmental reasons, need to be promoted in the child's gut are the ones that greedily feed on them. So it's basically a prebiotic, like a fertilizer for the organism – a probiotic is the actual organism."

While a great deal has been learned about the role and function of microbiota, their significance is still only beginning to be understood.

There are about 100 trillion of these organisms in our gut, making humans among the most complex ecosystems on the planet. To give you some perspective on what that number means, there are only about 27–37 trillion cells in the whole of the human body. This means that about three-quarters of the cells that make up you are not "you" at all, but organisms from a wide variety of other species. Kilograms of your bodyweight are made up of microbiota consisting of hundreds of different species. And as we know, they play key roles in digestion, metabolism and even in psychological wellbeing and mental health.

Humans are a grassland species, and if we think back to that diagram in the introduction (17), during all the visible parts of it once humans were born, they were immersed in an environment where they were surrounded by friendly bacteria (with only a fraction being not-so-friendly) that quickly socialized their immune systems into choosing the battles it really needed to win, and giving up on the kind of highly-strung housekeeping that rejects everything and flags it to "pursue and destroy". With allergies, the immune system keeps house as if it is vigorously polishing a thumbprint from a window and breaks the glass in the process.

Our immune system when we are born is a little like our brain. Our childhood lasts a long time (compared with many other species) because much of our neural circuitry is wired up after we are born. Rook explains that the immune system has, "got the hardware; it has got the software, but it needs information, just like the brain does." It's like a computer database, ready to sift and sort mountains of information, only waiting for the data to be keyed in.

We are born ready to learn about our environment. There is no point in having an immune system that is genetically hard-wired at birth to protect the body against things that may not be present in the environment you're born into. Much better to have one in which some of the components are innate and some are adaptive and so learn about the environment they are in and then concurrently develop protection against it.

The organ that works overtime in newborns to acquaint itself with the threats that the environment presents is the thymus. It is a small gland above the heart, and is a fraction of its size, but the thymus in babies (and young children) is proportionately much larger, as its work in the early years is to make friends with the body it is in and its environment.

Once born, we find ourselves in the Anthropocene surrounded by concrete and brick and glass, in clinical environments designed to keep bacteria at bay, regardless of their intentions. For millions of years we drank from streams, but now we drink chlorinated water, which means that we do not encounter previously common *saprophytic mycobacteria*. These bacteria are just one among many of the organisms that we have fewer opportunities to meet in modern life, with parasitic worms, lactobacilli also on the list. All of which are thought historically to have been important for our immune systems.

People who suffer severe allergies have substantially less diverse (and less dense) flora in their gut. It's believed that during our developmental years, we do not encounter the right microbiota and instead of brushing up against all kinds of bacteria on the grasslands of prehistory, we meet a bacterial population of unhelpful homogeneity, like a microbial version of IKEA. Instead of the complex diversity of a forest, it's more like an airless airport lounge.

Our brains rest and restore themselves best when they are taken home to the grasslands. But research suggests that our immune systems also need to be challenged by the kinds of biodiversity that can only be found in the wild. It seems that our immune systems are reluctant to learn substantial amounts of new information once the developmental window closes during early childhood.

During the 19th century, Charles Blackley observed that people who grew up on farms suffered fewer allergies.^ He was right and his observation has found a cogent explanation in the Old Friends hypothesis.

OUTDOOR VERSUS INDOOR

The World Health Organization believes that 7 million premature deaths annually are linked to pollution – that's 1 in 8 deaths globally.

There has been rising concern recently about indoor pollution. In the West, most people's homes will have some – and there has been stacks of legislation since the Victorian period to protect those in the workplace (thank you, Dr Ramazzini). But there are still some significant risks.

The most common sources are household appliances (gas cookers produce NO_2), poor ventilation (which can prevent the dispersal of radon or carbon monoxide) and toxic fibres used in items such as building materials. Smoke is an obvious one, as well as mould spores.

In almost all cases, the problems caused by indoor air pollution disappear with the wave of a fan. If the air is moving, pollutants are unlikely to bother us as they drift away into the ether. But what if 5 million households are all ventilating their polluted homes?

These problems are relatively easy to deal with, but they still contribute to one of the biggest causes of death globally. Of the 7 million deaths caused by air pollution, 3.3 million in southeast Asia and the Pacific region alone are thought to have been caused by indoor pollution (the world total is 4 million). The demographic means that predominantly poorer women and children are most affected.

The numbers concerning pollution are never going to be accurate or perfect. Who among the people counted was not exposed to both indoor and outdoor pollution? We are, most of us, consumers and producers of both indoor and outdoor pollutants. It makes sense then that the numbers are further confirmed by a study published in *Nature* in 2015, which estimated that the indoor air pollution caused by burning biomass fuels in China and India went on to cause 760,000 outdoor air pollution deaths.^

The numbers, like the air they are describing, are not "clean". They do signal that these places suffer terribly as a result of unsustainable practices in terms of transport, household energy supply and the management of waste.

Just under half of the people on the planet use open fires and basic stoves burning wood, dung, coal and the waste from crops. The number of people who do not have easy access to electricity is 1.2 billion; they turn instead to solutions such as kerosene lamps which emit masses of fine particulate matter into the household. The problem seems less connected to transport and car use than to outright poverty, where the health gap starts to look a lot like a wealth gap.

There is little doubt, too, that air quality is the chief driver behind the rise of chronic obstructive pulmonary disease (COPD). This debilitating condition is affecting ever more of us as the years go by. The World Health Organization estimates the number as 384 million globally – and by 2030 it is set to climb in the rankings to the third most common cause of death.^ Because the disease develops over months and years, its initial detection can be difficult, with 80–95 percent of sufferers unaware that they have it. As with so many conditions, COPD is a gateway pathology that leads to other problems. As the patient's ability to breathe is constricted over such a long period, changes in their lifestyle, such as a reduction in the range, duration and intensity of exercise go unnoticed, which in turn raises the risk of suffering from most of the other big killers.

Air pollution is in the frame as a cause, but the science isn't quite there. Exposure to workplace pollutants such as cadmium dust and fumes, grain and flour particles, silica dust, welding fumes, isocyanates and coal dust are known causes. There is about a one percent chance that you will have a genetic inclination to COPD, but that does not mean that the condition will progress. The two biggest factors are having a relative with COPD and smoking. COPD, with its associated causes, is the essence of an Anthropocene pathology.

NO_X & GREY FEVER

Mortality rates for asthma are much worse in low and middle income countries. The World Health Organization Mortality Database of 2014 reports Mauritius, Fiji, the Philippines and South Africa as the worst affected. In the UK more than 1,200 people die of asthma every year (about one person every eight hours). In Australia, the figure is worse, with one person in nine suffering from it (and rates of asthma are almost twice as high among indigenous Australians as non-indigenous Australians).^ For many it is seen as a minor inconvenience; for a few hundred million it is a life-threatening disability, but for hundreds of thousands each year, it's deadly.

I am among the five million people in the UK who live with asthma. I am on the more risky end of the scale – I have to medicate for it every day without fail. Like most asthmatics who die as the result of an attack, I believe that I have never had one serious enough to be life-threatening. I believe it to be under control, but it is precisely because of this belief that people die from it. They get comfortable with it, they think they know its signs, then with a little slip as easy as forgetting their keys when they leave the house, they suddenly find themselves unmedicated, with no access to immediate help – and they're for it.

About a year ago, I went out for a run one day, aged 47, and returned a decrepit octogenarian. I can remember the run in forensic detail. At first,

I was nonplussed that my regular running pace seemed completely out of reach. I had to keep stopping to catch my breath and the symptoms worsened as my panic rose. The symptoms I was experiencing were completely new to me. I thought I might have had a stroke or minor heart attack because I wasn't wheezing. I knew what asthma attacks sounded like, felt like – even tasted like – and this wasn't one. Toward the end I was going so slowly that a skilled anthropologist would have struggled to recognize my movements as running. Even though I had admitted defeat and turned back to the house, I pulled and I pulled at the air for breath, but I felt as though the air had been diluted; instead of deep blue air, this was thin, greasy and stagnant.

Later that day, a BBC newscaster warned asthma and hay fever sufferers to stay indoors because of high pollution and pollen levels. On Twitter, I saw fellow runners asking if others were struggling to breathe in London. Something had obviously gone very wrong.

The following day my doctor prescribed me oral steroids, but for the first time in my life they didn't really work. Steroids are a sort of inflammation carpet bomb – but the allergens I was reacting to were too potent for the steroids to have any effect. The doctor warned me off exercise for a fortnight, which didn't sound like a good idea to me. "Not even in a gym?" I asked. But she told me, "Our priority at the moment is to keep you out of hospital."

The most likely candidate was a new kind of weaponized air pollution: grey fever. Also referred to as urban hay fever, it is the result of engaging with higher levels of allergens and caused by their mode of delivery.

Diesel is the worst offender when it comes to air pollution. Diesel ownership has been exacerbated by the fact that so many drivers in the last decade were encouraged to buy cars with diesel engines because they had been assured that they were cleaner than petrol engines. In 2015 the story broke that many car manufacturers' diesel engines performed far worse on the roads than they reportedly did in laboratory tests. The scandal that followed revealed that several major manufacturers had deliberately misled the public, particularly over the levels of emitted nitric oxide and nitrogen dioxide (both generically known as NO_x). And it wasn't that the tests proved to be inaccurate, some cars were found to be chucking out more than 12 times the EU's maximum NO_x levels.

What is so loathsome about NO_x is not just that it's poisonous, but that it has no scent or taste and is invisible. It has been linked with memory loss, early-onset dementia, emphysema and chronic bronchitis. In the UK alone, it is thought to be responsible for up to 40,000 deaths a year. That is more than the sum of deaths from breast and lung cancer.

Pollen normally floats a little, depending on its weight and mass, then lands on the ground (though some has been tracked making its way to other countries across seas). On the ground, unless you are a bloodhound, you are relatively unlikely to inhale it. Pollen's other fate is to be swept into the upper atmosphere where it never bothers us again.

Atmospheric conditions can interfere with this process, sometimes with deadly results. Thunderstorm asthma is a recent phenomenon that has occurred in several countries over the last couple of decades, especially Australia. Its name connects it to severe weather events of the kind that are becoming more common in the Anthropocene due to global warming.

On an average hot summer's day, the pollen count is likely to be quite high. But with an oncoming storm, the air temperature suddenly drops, creating a lot of wind through air-pressure differentials. The fallen pollen swells up from the ground in the wind and is whipped about with a ferocity that ruptures the grains.

When most pollen is inhaled, it becomes caught on hairs or the mucal membranes of the nose and throat (hence the symptoms of a runny nose and an itchy cough; the pollen allergy leads to local inflammation). But broken pollen is much finer and can make its way down into the lungs, delivering a massive allergen load. The inflammation caused there by an allergic reaction can lead to an asthma attack – even a potentially fatal one. And for many of those involved, this might be their first attack. A storm in Melbourne in 2016 left at least four people dead, with many hospitalized.

Grey fever is a little less dramatic, but more dangerous because it affects much larger areas, and particularly the most populated ones. Settled pollen is whipped up by moving vehicles, which not only circulates it, but can rupture it, too. What makes this worse than normal hay fever is that the pollen, which can be very fine by this point, can attach itself to the black carbons and particulates emitted by cars (particularly those with diesel engines). The pollen cannot escape into the upper atmosphere, so the polluted air is heavier and does not disperse very easily. The pollen and its new best friends the particulates, churn and churn like a washing machine filled with poison. The particulates also make the pollen fragments more adhesive so they are more difficult to exhale.

Despite the fact that the pollen count in the UK has been consistently dropping, allergic symptoms in the population are worsening. There are several factors that might explain this, but the most likely is that the particulates to which the pollen is attached are not innocent bystanders – they are also highly noxious. Some are so small that they can penetrate the lining of the lungs and enter the bloodstream where they become long-staying guests.

One of the worst places on the planet that you might experience all these symptoms together is London's Oxford Street, where it is hard for even the olfactorily-challenged like myself not to notice the stench of diesel as the bactrian humps of London's red buses stretch into the distance.

In 2012 there were press reports about the surrounding streets being hosed down at night with a chemical glue (calcium magnesium acetate) that carbon emissions stuck to like a kind of fly paper.^

NO_2 poisoning is gruesome. It is caused when people reach a threshold of ingestion; farmers, firefighters and transport workers are particularly at risk. When inhaled NO_2 hydrolyzes to nitric acid in the lungs and goes on to cause fever, rapid breathing and heart rate, and severe shortness of breath – a perfect storm for asthmatics.

As many as 9,500 people die in London alone as a result of both NO_2 poisoning and that of PM2.5s (particulate matter measuring less than 2.5 millionths of a metre in diameter).^ If you cut a human hair, you would be able to balance about 100 of these on the exposed edge of it (as 2.5μm is their maximum size). The size of the particles is important because that is what directly links them to health problems such as premature death, allergic reactions, lung problems and cardiovascular diseases which lead to hospital admissions.

The particles are so small that they are suspended in mid-air and a governmental report by the Air Quality Expert Group in 2012 stated that they come from a range of sources.^

- Vehicle exhaust emissions from diesel engines
- Tyre and brake wear
- Road abrasion
- Road dust redistribution
- Dust emissions from construction
- Demolition

- Quarrying
- Mineral handling
- Other industrial and agricultural processes
- Domestic and commercial cooking
- Small-scale waste burning and bonfires
- Domestic wood burning
- Biogenic emissions
- Emissions of NH_3 (ammonia) from agriculture
- Off-road construction and industrial machinery
- Emissions of SO_2 and NO_x from shipping

The effects of pollution worsen during long dry spells because rain rinses PM2.5s from the air. In the winter, too, during a high-pressure chill local emissions from traffic and power plants can build up for days before being dispersed by changing weather. Data from DEFRA (Department for Environment, Food & Rural Affairs) show that the worst affected areas are all the places you've heard of: London, Birmingham, Manchester, Bristol, Southampton and many others.^ The map DEFRA produced illustrating the spatial distribution of PM2.5s looks exactly like a road map, with all the cities highlighted, and the rich network of motorways and A-roads that connect them tinted as areas of high pollution.

The only clean areas are the 13 little lungs of Britain: its national parks. In the US, most of the east coast is clean, with California and Seattle vying for prominence on the west. In India the major conurbations in the northern quadrant (such as New Delhi) are reducing the life expectancy of about 600 million people by an average of 3.2 years (some two billion life years in total). The southern quadrant of the country leads the way for indoor particulate pollution.

A number of UK cities are expected to continue to exceed EU limits on air quality until 2030. In the UK, 8 percent of deaths every year are caused by the air you are breathing as you read this. Considering that we are one of the "developed" nations, these figures put the UK slightly ahead of the world trend. A study by more than 50 scientists published in *The Lancet* in 2015^ ranked air pollution as eleventh in the world for its influence upon something that epidemiologists call disability-adjusted life years. These are calculated by adding the years of life lost to years lost due to disability. These are fairly self-explanatory terms; the first is how many years of your life you lose to an early death; the second is years lost due to the overwhelmingly poor quality of life brought on by the cause. This is not a good ranking for the UK, and it is one of the easiest for governments to tackle.

A recent report by the World Health Organization tells us that up to 92 percent of the world's population is living in a poisonous atmosphere.^ And even though smoking has seen a steep decrease in high-income nations in the last 20 years, levels of lung cancer are still high in densely-populated and more affluent parts of the world. Of the estimated 7 million people who died from air pollution in 2016, 21 percent died from pneumonia, 20 percent from stroke, 34 percent from heart disease, 19 percent from COPD, 7 percent from lung cancer (that's 490,000 worldwide).

Air quality matters. A trial published in *Psychological Science* in 2018 examined the ways in which air pollution negatively influenced ethical and criminal behaviour by raising levels of anxiety (particularly the production of cortisol).^ This was no speculative study; the researchers gathered data from 9,360 cities in seven crime categories (murder, rape, robbery, burglary, car theft, theft and assault). They found, "air pollution positively predicted incidents of every crime category in all models."

As if by magic, while I was working on this chapter my asthma worsened dramatically for the second time. The following day it was as if there was a sea fret in south London. The air was thick, heavy and oddly warm for late autumn. The heat, just like the pollen and PM2.5s, was unable to escape. Again, I had none of the usual symptoms: I had no wheezeing and my inhalers seemed to have little effect, despite my quadrupling the dose. I developed a cough. Even walking left me out of breath; there was no chance of going for a run. A day or two later, London's mayor, Sadiq Khan, tweeted that he was issuing his seventh pollution alert in just over a year.

I thought about writing to the mayor and putting myself up for a job. I could be London's canary, like the birds coal miners once used, with my especially sensitive early-warning system lungs able to detect pollution before anyone else could.

In October 2017, it was revealed that the entire capital had exceeded World Health Organization guidelines for levels of PM2.5s.

Today Londoners are coughing and choking as much as they ever have done throughout the last thousand years. The only difference is that this time those in power don't really seem to care.

On the one hand, we are badgered to walk or cycle to work, to exercise more; while on the other the quality of the air that we must suspend ourselves in to do these things is damaging our health. Governments can ask all they like, but as long as people have their cars – and they are much nicer and safer environments than walking or running beside a road – they will have an impossible job in coaxing the drivers to get out of them. Action is required, not persuasion.

The solution is both extremely simple and complicated. Cut emissions from vehicles. Make using cars a less attractive prospect for drivers, rather than just punishing them with charges. Put in place a better functioning public transport system designed to meet the uptake in use. Make cities safer and more walkable and cycleable. Cleaner air will result. Easy!

After that initial bout of asthma on that summer run, my asthma improved about a fortnight later, and about six weeks after that I was nearly back to normal. But the improvement plateaued, and so much time went by that I began to believe I had mistaken convalescence for aging, because the truth is, I have never felt the same since. I have never, to this day, got my fitness back.

Why is nothing likely to be done about this? The global automotive industry is worth approximately $800 billion a year and 80 million cars are made every year. The industry is planning to scale up production to 100 million by 2020.

In the UK, the Office for Budget Responsibility predicts that the Exchequer will raise £28.2 billion in fuel duty this year. The car industry in the UK also sells about 2.7 million cars annually; a conservative estimate of the VAT raised from this is at least a further £7.5 billion. Curbing car production would put a £35 billion hole in the Exchequer's pocket.^

In this festival of idiocy, the British government also recently removed one of the key incentives for owning a cleaner vehicle by hiking road tax levies on the least-polluting cars and cutting it on the most poisonous. And the injudicious policy decisions continue with the fact that they recently spent £370,000 defending and losing two actions brought by Client Earth that aimed to expose the fact that the government had done nothing to tackle poor air quality. For now, the current (and future) government has to decide whether to continue choking its taxpayers or choke the economies creating the problem – which would lose them votes.

The environment may have changed, but many of the problems relating to air quality in the 19th century persist today – it's just that the numbers involved are greater. And, unlike the Victorians, we persist in our behaviours in the full knowledge of what pollution is doing not just to our generation, but also to the next. A report published in *JAMA Psychiatry* in 2015 presented some shocking evidence of this. The researchers were investigating the potential damage caused by air pollution to unborn children. Tracking them over nearly ten years, they found that exposure to a wide range of "air pollutants contributes to slower processing speed [during intelligence testing], attention-deficit/hyperactivity disorder symptoms ... and conduct disorder problems" throughout the cohort they looked at.^

Our environment, and the air in it, is changing our children's bodies and brains before they even take in a lungful of soot-flecked air.

WINDING BACK

1 Stay connected

It is crucial that as a species we maintain as much contact with green spaces and the natural environment as we can, for hundreds of reasons to do with our biology and psychology, but also for the ecology within our bodies. Consuming a varied diet with many different fruits and vegetables helps to maintain biodiversity in the gut. This is easy to recommend, but the science is also telling us to do more challenging things such as minimizing the use of antibiotics, especially during pregnancy. The science is also telling us to avoid limiting the transmission of maternal microbiota to an infant, through C-sections or by not breast-feeding. (As a man who has never had to think about whether or not to breast feed, it is not easy to recommend this, especially when so many societies are not set up to encourage it.)

2 Feed your gut as well as yourself

Microbiota are stimulated by the kinds of foods we eat, so we should eat the kinds of foods that stimulate microbiota. Just as our height can only spurt at given moments during our development, the window of opportunity to train our immune systems is small, and anyone who can read this is already too old to fully retrain it. That does not mean, though, that microbiota do not matter – far from it. Eat diverse foods, especially fresh fruits and vegetables – fibrous foods such as broccoli and other cruciferous vegetables (kale, cabbage and cauliflower) are good because we can't use the fibre in them, but our gut bacteria can (in the same way as the polysaccharides in breast milk that we can't digest). Fermented foods are good, too: foods such as tempeh, miso, yogurt and kefir. Probiotics can reduce insulin, triglycerides and cholesterol levels in those who are obese, which means they can reduce risk factors for a number of diseases linked with obesity, including heart disease and type 2 diabetes.

Microbiota also drive appetite. The science is still in its early stages (with experiments being carried out on populations of fruit flies, for example), but it is already showing that species of bacteria respond to the foods we eat by growing or shrinking their population. When we eat, they do, and they secrete substances, activate genes and help us to absorb nutrients. They also communicate with our central nervous system and our brain. How they do this, researchers don't yet know, but it looks as though over the millions of years that the microbiota have lived inside humans, we have co-evolved as communicative hosts. They drive appetite so we give them what they want. And when we create a hostile environment for them, our bodies respond with symptoms of depression, anxiety – even high blood pressure.

3 Be/stay active

If estimates for the dip in the nutiritional content of our foods are correct, we may be driven toward compensatory eating. But gaining weight is not the forgone result of this – it is just a mathematical model. Change the numbers by incorporating a little extra movement into your day. Exercise is widely viewed by governments and scientists alike as a miracle cure, and they are absolutely right. It fights all the big killers, even some cancers and neurodegenerative diseases. We only have brains because we move, and plants don't need them because they are static. Don't be a plant.

4 Be a 5-a-day sceptic

Like the Active 10, this is not a bad goal, but the key to healthy microbiota is variety, so mix it up where you can.

The following are signs that you may be nutrient deficient:

- Appearing unseasonably pale; iron deficiency can lead to smaller and/or fewer red bloods cells. Paleness around the eyes, the mouth or gums are the giveaways. See your doctor, but eat lentils, beef, spinach, beans and broccoli to tide you over till you do.

- Thinning or brittle hair has two common causes, a protein or vitamin C deficiency. (If you've ever had this, you're in good company, Lord Byron was a serial crash-dieter and suffered hair loss and discolouration). Your hair can also turn grey prematurely if you are deficient in copper or vitamin D. Tuck into some hazelnuts or almonds for extra copper; dairy products, eggs and oily fish are good for vitamin D, but nothing is better than a sunny holiday. In the US, about 42 percent of people may be D deficient (74 percent of the elderly; and as many as 82 percent of people with darker skin).

- The calcium content of fruit and veg is dropping and it's needed in every cell in our body. If you do not get sufficient calcium from your diet, it is released from your bones, which can lead to rickets and osteoporosis. Those emerald vegetables have lots of calcium (broccoli, kale and so on) as do dairy products. A can of sardines will give you 44 percent of your RDI (reference daily intake; the daily intake level of a mineral required to meet the requirements for 97–98 percent of healthy individuals in the US), and recent research proves that your microbiota will love it, too.

5 Lift your pelvis

If you have an anterior pelvic tilt, more regular exercise is the answer for lifting it. There's no reason to focus specifically on your core; pretty much any exercise that works your abdomen in some way will do. Many people believe that their spine is more stable when it is straight and "engaged". This is not true. For normal lifting, the gentle S-shape of the spine should be there; that is its original shape so that is how it performs best.

6 Don't eat dirt

Despite what you might think after reading this chapter: don't eat soil.

7 Write to your MP

Ask why they are doing nothing about air quality, and why nothing is being done to build bigger and better green spaces.

8 As an absolute minimum do 10,000 steps

Ten thousand steps probably cover about 7km (4.3 miles). The problem with 10,000 is that it is a completely arbitrary number. A meta-analysis published in 2011 produced head-scratching results. "It appears that healthy adults can take anywhere between approximately 4,000 and 18,000 steps/day, and that 10,000 steps/day is a reasonable target for healthy adults."^ The variance in numbers emerged because the ranges of reported exercise intensity during the 4,000–18,000 steps diverged greatly. A squash game, for example, may only encompass 4,000 steps, but the cardiovascular benefits derived from it are comparatively enormous. I suspect we should aim for the kind of mileage that people have covered for two million years or more years: 8–4.5km (5–9 miles), about 12–18,000 steps per day. If a nice round number is what you are after, 100,000 steps per week is a bit more hunter-gatherer.

9 Try to go barefoot more often

Shoes encourage us to take longer strides. Cushioned heels encourage a harder foot strike so that being without shoes seems less comfortable because we are performing a walking style that we have developed while shod. Try to walk or work out (including doing weights) either barefoot, or in an unsupportive shoe that capitulates to your foot, so the intrinsic muscles in your feet are allowed to respond to the weights and exercise in the same way you expect your body to.

10 Sign up to Gary Ward's online movement class (findingcentre.co.uk)

11 Beware the Wolff

Wolff's Law (developed by the German anatomist and surgeon Julius Wolff in the 19th century) states that bone in a healthy person or animal will adapt to the loads under which it is placed. Our bodies' muscles, tendons and ligaments adapt to loads, or the lack of them, so move them and use them if you don't want to lose the use of them.

12 Move it or lose it

Without a doubt the best way to avoid back pain is through exercise, creating a body that tolerates motion and loading.

It is a misconception that movement makes back pain worse; research says the opposite. The same goes for exercise. Under worked muscles are very easy to injure but trained muscles less so. Professor Mike Adams explains, "what doesn't kill you, makes you stronger. All tissues respond to mechanical loading. When it comes to bone, loading makes the tissue stiffer and stronger. Cartilage is the same, but it is very slow and can take years to respond. Muscle is fast to respond."

A study published in 2018 suggested that astronauts who experience long bouts of microgravity and consequently a lack of loading in the spine, can suffer spinal stiffness, disc swelling and muscular atrophy.^ Unloading the spine in this way is associated with back pain. The recommendation is to lift or load regularly; make it part of everyday activity. Rowers who constantly load their spine at full flexion, if they are experienced, have rates of back injury that are about the same as middle-aged sedentary workers. Professor Mike Adams says: "Olympic athletes often have a lot of spinal pathology (to bones and discs), but no more pain than an average woman or man." The action of rowing is about the most stressful activity a spine can undertake, yet with some training little injury results. Beware, though; sudden increases in any kind of training are strongly associated with injury.^

13 Pain does not equal damage

This is a difficult one for sufferers to understand, as strong pain often seems to signal the opposite. But research increasingly shows that social and cultural influences (as well as elements such as symptom duration and patient beliefs) are more important in quelling or amplifying pain than anything that might appear on a scan.

14 Consider a scan with caution

Going for a scan fails to source the cause of back pain in about 99 percent of cases. It won't be long before muscular problems are identifiable on an MRI – but in most cases, scanning wouldn't be worth the expense.

15 Get a well-fitting bra

Large breasts can put a strain on the spine. The underwiring of a well-fitting bra should stay next to your skin when you raise your arms above your head. If it doesn't, it may be time for a refit.

16 Make a stand against sitting and consider a different working environment

Any change to your working environment which encourages you to move more is good. A standing desk might be a good idea. You will probably be more active using one, but make sure it is adjustable so you can stand or sit when it suits you. Also bear in mind that there is no "natural" behaviour that I am aware of that mimics standing in a static pose in one place for extended periods. Early humans peered over the long grass, but did not stand like statues. Stand for as long as you want, then sit for a rest.

17 **Help prevent allergies emerging in the young**
Get a dog. Dogs have been proven to challenge the hyper-cleanliness of modern life (in a good way) with greater ease than many interventions. Their need for frequent walks is very useful to get us outside and walking briskly at least twice a day.

18 **Protect yourself by showering before bed**
Showering helps to rinse away allergens and PM2.5s that have settled in your clothes and hair through the day. That way you won't take them to bed and grind them into your pillows so your eyes itch every night until you wash the bedclothes. Also, use a top sheet. Put a thin cover on your bed during the day and carefully remove it at night; this will prevent any allergens that have settled there from affecting your sleep. Any wide-brimmed hat will help to keep allergens away from your eyes and nose during the day, as will wraparound sunglasses, which will be an effective shield for your eyes.

…color:#353535;mso-ansi-language:EN-US'>**2025** </span></b><span lang=EN-US style='font-
family:AppleSystemUIFont;mso-bidi-font-family:AppleSystemUIFont;color:#353535;mso-ansi-language:EN-US'><o:p></o:p><
span></p><p class=MsoNormal style='mso-layout-grid-align:none;text-autospace:none'><spanlang=EN-US style='font-
family:AppleSystemUIFont;mso-bidi-font-family:AppleSystemUIFont;color:#353535;mso-ansi-language:EN-US'>**Cure for the**
common cold<o:p></o:p></span></p>…color:#353535;mso-ansi-language:EN-US'>**2027**</span></b><span lang=EN-
USstyle='font-family:AppleSystemUIFont;mso-bidi-font-family:AppleSystemUIFont;color:#353535;mso-ansi-language:EN-
US'><o:p></o:p></span></p><p class=MsoNormal style='mso-layout-grid-align:none;text-autospace:none'><spanlang=EN-
US style='font-family:AppleSystemUIFont;mso-bidi-font-family:AppleSystemUIFont;color:#353535;m
so-ansi-language:EN-US'>**First disease eradicated by gene editing**<o:p></o:p></span></p>…color:#353535;mso-
ansi-language:EN-US'>**2032** </span></b><span lang=EN-USstyle='font-family:AppleSystemUIFont;mso-bidi-font-family:Appl
SystemUIFont;color:#353535;mso-ansi-language:EN-US'><o:p></o:p></span></p><p class=MsoNormal style='mso-layout
grid-align:none;text-autospace:none'><spanlang=EN-US style='font-family:AppleSystemUIFont;mso-bidi-font-family:AppleSys
temUIFont;color:#353535;mso-ansi-language:EN-US'>**Driverless cars become the norm in most countries**<o:p></
o:p></span></p><p class=MsoNormal style='mso-layout-grid-align:none;text-autospace:none'><b><spanlang=EN-US
style='font family:AppleSystemUIFontBold;mso-bidi-fontfamily:AppleSystemUIFontBold;color:#353535;mso-ansi-language:EN-
US'>**2035** </span></b><span lang=EN-US style='font-family:AppleSystemUIFont;mso-bidi-font-family:AppleSystemUIF
ont;color:#353535;mso-ansi-language:EN-US'><o:p></o:p></span></p><p class=MsoNormal style='mso-layout-grid-
align:none;text-autospace:none'><spanlang=EN-US style='font-family:AppleSystemUIFont;mso-bidi-font-family:AppleS
ystemUIFont;color:#353535;mso-ansi-language:EN-US'>**10 percent of humans have Type 2 diabetes; gut flora**
transplants become common treatment for mental health issues, obesity and allergies<o:p></o:p></
span></p><p class=MsoNormal style='mso-layout-grid-align:none;text-autospace:none'><b><spanlang=EN-US style='font-
family:AppleSystemUIFontBold;mso-bidi-font-family:AppleSystemUIFontBold;color:#353535;mso-ansi-language:EN-US'>**2037**
</span></b><span lang=EN-USstyle='font-family:AppleSystemUIFont;mso-bidi-font-family:AppleSystemUIFont;color:#35
3535;mso-ansi-language:EN-US'><o:p></o:p></span></p><p class=MsoNormal style='mso-layout-grid-align:none;text-
autospace:none'><spanlang=EN-US style='font-family:AppleSystemUIFont;mso-bidi-font-family:AppleSystemUIFont;color:#35
3535;mso-ansi-language:EN-US'>**Implant technology that monitors health is rolled out to the entire population**
- patients no longer relied upon to report symptoms<o:p></o:p></span></p>…color:#353535;mso-ansi-
language:EN-US'>**2040** </span></b><span lang=EN-USstyle='font-family:AppleSystemUIFont;mso-bidi-font-family:AppleSyst
emUIFont;color:#353535;mso-ansi-language:EN-US'><o:p></o:p></span></p><p class=MsoNormal style='mso-layout-gric
align:none;text-autospace:none'><spanlang=EN-US style='font-family:AppleSystemUIFont;mso-bidi-font-family:AppleSystemU
IFont;color:#353535;mso-ansi-language:EN-US'>**New strain of swine flu kills 300 million worldwide**<o:p></o:p></
span></p><p class=MsoNormal style='mso-layout-grid-align:none;text-autospace:none'><b><spanlang=EN-US style='font-
family:AppleSystemUIFontBold;mso-bidi-font-family:AppleSystemUIFontBold;color:#353535;mso-ansi-language:EN-US'>**2045**
</span></b><span lang=EN-USstyle='font-family:AppleSystemUIFont;mso-bidi-font-family:AppleSystemUIFont;color:#35
3535;mso-ansi-language:EN-US'><o:p></o:p></span></p><p class=MsoNormal style='mso-layout-grid-align:none;text-
autospace:none'><spanlang=EN-USstyle='fontfamily:AppleSystemUIFont;mso-bidi-font-family:AppleSystemUIFont;-
color:#353535;mso-ansi-language:EN-US'>**Common use of AI in the workplace results in mass unemployment,**
with those in work suffering high levels of stress, mental health issues proliferate in both groups<o:p></
o:p></span></p><p class=MsoNormal style='mso-layout-grid-align:none;text-autospace:none'><b><spanlang=EN-US
style='font-family:AppleSystemUIFontBold;mso-bidi-font-family:AppleSystemUIFontBold;color:#353535;mso-ansi-language:EN-
US'>**2055** </span></b><span lang=EN-US style='font-family:AppleSystemUIFont;mso-bidi-font-family:AppleSystemUIF
ont;color:#353535;mso-ansi-language:EN-US'><o:p></o:p></span></p><p class=MsoNormal style='mso-layout-grid-
align:none;text-autospace:none'><spanlang=EN-US style='font-family:AppleSystemUIFont;mso-bidi-font-family:AppleSystem
UIFont;color:#353535;mso-ansi-language:EN-US'>**Post-antibiotic era begins in which resistant bacteria flourish in**
new warmer climates posing constant risk to health <o:p></o:p></span></p><p class=MsoNormal style='mso-
layout-grid-align:none;text-autospace:none'><b><spanlang=EN-US style='font-family:AppleSystemUIFontBold;mso-bidi-font-f
mily:AppleSystemUIFontBold;color:#353535;mso-ansi-language:EN-US'>**2057** </span></b><span lang=EN-USstyle='font-
family:AppleSystemUIFont;mso-bidi-font-family:AppleSystemUIFont;color:#353535;mso-ansi-language:EN-US'><o:p></o:p><
span></p><p class=MsoNormal style='mso-layout-grid-align:none;text-autospace:none'><spanlang=EN-US style='font-
family:AppleSystemUIFont;mso-bidi-font-family:AppleSystemUIFont;color:#353535;mso-ansi-language:EN-US'>**Sardinian**
woman lives to 150 years old; first fully-automated and mobile weapon systems launch to police the
streets and go into combat zones<o:p></o:p></span></p>color:#353535;mso-ansi-language:EN-US'>**2070** </
span></b><span lang=EN-USstyle='font-family:AppleSystemUIFont;mso-bidi-font-family:AppleSystemUIFont;color:#353
535;mso-ansi-language:EN-US'><o:p></o:p></span></p><p class=MsoNormal style='mso-layout-grid-align:none;text-
autospace:none'><spanlang=EN-US style='font-family:AppleSystemUIFont;mso-bidi-font-family:AppleSystemUIFont
;color:#353535;mso-ansi-language:EN-US'>**Orbital ecology is so crowded that hundreds of thousands of**
satellites are destroyed in collisions - digital communication becomes a premium commodity </
span><b><spanlang=EN-US style='font-family:AppleSystemUIFontBold;mso-bidi-font-family:AppleSystemUIFontBol
d;color:#353535;mso-ansi-language:EN-US'><o:p></o:p></span></b></p>color:#353535;mso-ansi-language:EN-
US'>**2075** </span></b><span lang=EN-USstyle='font-family:AppleSystemUIFont;mso-bidi-font-family:AppleSystemUIF
ont;color:#353535;mso-ansi-language:EN-US'><o:p></o:p></span></p><p class=MsoNormal style='mso-layout-grid-
align:none;text-autospace:none'><spanlang=EN-US style='font-family:AppleSystemUIFont;mso-bidi-font-family:AppleSyst
emUIFont;color:#353535;mso-ansi-language:EN-US'>**Human population crests 10 billion**<o:p></o:p></span></
p><p class=MsoNormal style='mso-layout-grid-align:none;text-autospace:none'><b><spanlang=EN-US style='font-
family:AppleSystemUIFontBold;mso-bidi-font-family:AppleSystemUIFontBold;color:#353535;mso-ansi-language:EN-
US'>**2100**</span></b><span lang=EN-USstyle='font-family:AppleSystemUIFont;mso-bidi-font-family:AppleSystemUIFont;col
or:#353535;mso-ansi-language:EN-US'><o:p></o:p></span></p> class=MsoNormal style='mso-layout-grid-align:none;text-
autospace:none'><spanlang=EN-US style='font-family:AppleSystemUIFont;mso-bidi-font-family:AppleSystemUIFont;color:#35
3535;mso-ansi-language:EN-US'>**first humans experiment with complete conversion to neuro-hypertext**<o:p><
o:p></span></p>…

Part V
The Future

HOMO SAPIENS INEPTUS

But man is the unnatural animal, the rebel child of nature, and more and more does he turn himself against the hard and fitful hand that reared him.

H G Wells, *Modern Utopia* (1905)

We began on our toes, exploring the curves of our arches and the adipose tissue on our heels. It seems a fitting end to conclude at our fingertips, those parts of our anatomy that now swipe and touch and type their way through modern life.

Like our feet, they started as fins. They are pentadactylic – and no one has yet come up with a convincing explanation as to why so many living things have five digits, from starfish, whales and kangaroos to primates. Horses and deer? Hooved quadrupeds basically run on their fingernails with their feet being an evolutionary elongation of the middle finger. Dogs look as though they only have four toes, but their fifth is hidden away a little further up their legs; it's called a dewclaw.

Our hands are also like our feet in so far as it is not unheard of for people to sport an occasional extra digit. Both structures possess nearly the same number of bones (with some believing that the talus bone is a fused version of two separate bones). Our hands and feet each have four kinds of arch muscles (remarkable given their different functions). The hand has fewer joints, but more muscles allow greater range and complexity of movement. This is why we find ourselves living in an information age for which three things are predominantly required: our eyes, our brains and the limbs that interact with all that data: hands.

Our hands and fingers have come a long way from supporting our body weight during quadrupedal motion and suspending our weight during brachiation. Over millions of years they adapted away from those functions to better support life on two feet.

We are not in the least unique as a species in our possession of opposable thumbs, but only humans have thumbs which are both strong enough and sufficiently divergent to have the thick muscles that give us unusual power blended with the kinds of cognitive processing that allow us to employ that grip to fire a bow and arrow, accurately propel a spear or hold a phone and with same hand text "On my way – 10 mins".

Hands are about as important to our evolution as sweat and brain size.^ So many parts of our anatomy are adapted to hunting and the hands are no exception. One of their most crucial roles in our evolution was in tool making and use, but also in the refinement of the precision of movement. While our feet are grounded, our hands are always opening into empty space. They are as patient as they are thirsty for the touch of another's skin or a phone. They spend much of their time not knowing, feeling their way in the dark. And they do everything.

Nearly all that our species has achieved or made has been with our hands. They have accomplished everything from picking up a stick to repairing parts of a structure on a space station. These two organs transformed corporeality into reality, biology into society, stones into tools, time into history.

Biology became history when we learned to paint, to write and to carve. There may as well have been magical lightning forking from our fingers for all the effect manual dexterity had on our evolution. Instead of relying on biological connections to share sentiments and a sense of social cohesion, instead of needing to touch, groom or have sexual intercourse to maintain group dynamics, the hand made it possible to achieve things at a much greater distance than an arm's length because it was the source of communication through gesture and sign. But gesture and sign alone are limited technologies. Where the hand had been used to fidget, caress and occasionally fight, it also unwittingly placed limitations on the kind of society that could be made by hand. For societies to grow beyond the size of a social group into, say, a city, other technologies and modes of communication are necessary.

CHAPTER 10

HANDS & A DIGITAL REVOLUTION

Sherlock Holmes' quick eye took in my occupation, and he shook his head with a smile as he noticed my questioning glances. "Beyond the obvious facts that he has at some time done manual labour, that he takes snuff, that he is a Freemason, that he has been in China, and that he has done a considerable amount of writing lately, I can deduce nothing else."

Mr Jabez Wilson started up in his chair, with his forefinger upon the paper, but his eyes upon my companion.

"How, in the name of good-fortune, did you know all that, Mr Holmes?" he asked. "How did you know, for example, that I did manual labour. It's as true as gospel, for I began as a ship's carpenter."

"Your hands, my dear sir. Your right hand is quite a size larger than your left. You have worked with it, and the muscles are more developed."

"The Red-Headed League", Arthur Conan Doyle

The workers have taken it into their heads that they, with their busy hands, are the necessary, and the rich capitalists, who do nothing, the surplus population.

The Condition of the Working Class in England, Friedrich Engels

At the core of our difference from other animals is supposedly our imagination, the ability to go beyond the present into the future and the past while simultaneously gathering moments and a community into ideas and a shared system of belief in how a society might or ought to function. With the emergence of art in our species, all these things speak in unity.

In Jill Cook's book, *Ice Age Art: Arrival of the Modern Mind*, she discusses the earliest known figurative sculpture.^ It is about 40,000 years old and was carved deep in the cold of the last European Ice Age. It was discovered in pieces at the back of a cave in southern Germany in the mid-20th century and was not assembled until several decades later. It stands about 30cm (12in) high and has been called The Lion Man.

It is recognizably an upright figure, but it is not a man wearing a mask, as the ears of the lion are pricked up as if aware of the Anthropocene human's attention. The arms are at rest beside the body, but the sculpture has them slightly separated from it. It has been estimated that the preparation of the sculpture would have been costly, not in terms of materials (as it is carved from mammoth ivory) but in human time.

Living in a harsh and demanding climate, this tribal group found it worthwhile to allow an individual to expend all their energy in the creation of this object. It is not just the object itself that is significant; the skill required to carve it was considerable, and this was clearly not the first example of this carver's craft. This single object represents hundreds of hours of work. Yet its figurative status as an object opens a door to a world long lost to us. It shows us a time and place in which a shared vision of how the world worked, how the living connected with the dead or the spirit world beyond. It shows us a shared belief system and a time in which shared belief became the principal means of belonging.

Carved by hand, with its surface smoothed, probably from being passed from hand to hand to hand, human dexterity made this, and the touch of countless others – probably belonging to many generations of the tribe – gave this object meaning, while it recursively gave meaning to the world around its people.

The Lion Man is a remarkable sight to behold because so much of us is already there, 40,000 years ago on the ice sheets of northern Europe. This was a society made by hand; social cohesion carved from a mammoth's tusk.

Just as handmade objects are highly individual, require skill and intelligence to produce and have limited possibilities for distribution, once knowledge, customs, ideas and ways of living can be transferred at distances greater than an arm's length (visually or through the written word) the possibilities become endless. Instead of having close social connections with only the people in your immediate environment, you can nowadays email someone in another country because you read some of their research, then go there, arrange to meet them, go home, write up the meeting and put it into your book.

The ability of the hand to create this distance between people and ideas seems more important in terms of our evolution than any individual body part might be. Once you have knowledge transfer, the possibilities for biological transcendence begin. This is the point at which the limitations of the body can be treated, enhanced or bypassed because intellect, whether individual or of a group, can work around a particular problem.

It all sounds rather idyllic, and perhaps it is. An idyll in which the body "boldly goes" beyond the supposed sociological limit of being connected to 150 people – basically, a medium-sized tribe. Our hands now do the work of this sociological connection, whether through text, WhatsApp, IM, Instagram, Facebook, Twitter or email – our fingertips are always at the rock face of communication, although we are fool enough to think that it is our mouths. And here there might emerge a peculiar mismatch disease – handmade, it seems.

The cognitive work of modern life is as easy to miss as it is substantial. It is difficult for city dwellers to imagine a world in which strangers are a novelty. For most, they are the majority of those we encounter and our way of coping is to behave in public as if we were alone in private.

The media environment in which we exist is one which is equally overwhelming. We evolved in small groups, gangs in which there was not a great deal of cognitive dissonance, without competing political world views, contradictory religions or alternative ways of being. As a result, we lack loyalty and the strong social bonds that would have been an inevitable outcome of tribal and manual relations, as would a world view shared by everyone in the group. We are, on the other hand, overwhelmed by choices and often conflicting ideas and views and we have to sift through great quantities of information.

In the Anthropocene, though, our hands seem just as taxed and overworked as our attentional faculties. There is always something for them to do. Our hands suffer from overuse; if we are not at a computer, we are texting, tapping and swiping our phones or tablets. In our spare time, we are gripping and wrestling with console controllers. If you don't share in these customs (and the numbers suggest there are about 33 million gamers in the UK alone), you are almost certainly surfing the net or using a computer all day, but despite this, our hands are weaker than ever.

Globally, internet use has unsurprisingly risen steeply since the beginning of the century. In 2000, the internet was still getting going with only about 2 billion people regularly online. In 2018, that figure has nearly doubled, with user-percentages well into the nineties in high-income nations. China alone accounts for about 750 million of internet users. The youngest users are the busiest: 70 percent of the world's youth are regularly online. In the UK in 2018, data tells us that 99 percent of people aged 16–34 are digital natives.^

Those are some busy habits. We are all using our hands nearly all the time and this has led to a new kind of digital morbidity known as RSIs (repetitive strain injuries). These are a whole range of pathologies associated with overuse. Their rise is connected to the shift in ecologies of labour – even in the Industrial Revolution a wide variety of work was still done by hand. No doubt people's digits were as exhausted as the rest of their bodies, but they were fulfilling a variety of activities which made their hands stronger, more tolerant and gave them better resistance to complaints. How long before our digits refuse to compute?

I clearly remember the excitement when the internet first arrived. I also remember getting my first computer in about 1995, and in 1998, getting dial-up installed at home. This technology has been digital in both senses of the word for decades. Today, it is still dominated by manual input, but the times are changing. We can hail Siri or Alexa from across the room, and as long as we ask, "What's the weather like today?" we are guaranteed a correct response. However, if we ask them to dim the lights to 60 percent, that becomes an approximation of their success rate.

When I bought my first computer some speech recognition software was included with it. It was clumsy and had to be trained for 4–5 hours, but it nearly worked. Since then, home computing processor power has increased by at least 3,000 percent, but speech recognition seems only marginally better. It has gone from about 97 percent accuracy to perhaps 97 percent; these sound like good numbers, but they mean an error per line and if you try to use them seriously they are frustrating. Nonetheless, the technology is en route and due to arrive soon.

HEY SIRI! WHAT'S RSI?

I had this dialogue with Siri.

"Hey, Siri. What's RSI?" I asked.

"What's our Açai? I didn't find any information over the next few months."

"What is RSI?"

"It's 9.53, good morning."

"No, what... is... R... S... I...?"

"I couldn't find any information for Our S I."

"WhatisRSI?"

"I couldn't find any information for rice eye."

"Whattttt. is. RSI?"

"OK. Here's some information."

Siri then loaded the Wikipedia page for the mismatch disease that is the scourge of occupational health practitioners. The above interaction is hardly likely to inspire fear that the machines are about to take over, but these Siris and Alexas will be nothing if not quick learners. The machines are coming, but in the meantime, it's still all down to our hands and they are not happy.

RSIs are a set of diseases that you, or someone you know has suffered from. A repetitive strain injury affects almost everyone at some point in their lives and can strike at any age. It affects the muscles, tendons and soft tissues of the hand and is associated with repetitive work, excessive vibration, compression and exertion (especially high-intensity activity) – usually the result of a sustained pose or awkward working position.

Before we had computers – and this is something every schoolchild is aware of, particularly at exam time – an RSI might simply have been writer's cramp from holding a pen too long, a condition that our Italian friend

Bernardo Ramazzini noticed in 1713 when he wrote about scribes who had the condition. Keyboards and mice are particular culprits, but typists and factory hands also suffered, along with musicians and sports people. In the past, RSIs were rare; today they cost UK employers in the region of £500 million a year in lost productivity – with six people a day leaving their jobs because of them.^

As many as three out of five sedentary workers in Sweden are affected, as well as up to 40 percent of Dutch university students. Across the board, women are affected by the disease more than men as noted in a report prepared forThe Robens Centre for Health Ergonomics European Institute of Health & Medical Sciences.^

It is not understood why particular RSIs develop in some people and not others; the associations are mighty strong, though. The symptoms include pain, aching or tenderness, stiffness, throbbing, tingling or numbness, weakness and cramp. Circulation is disrupted and the muscles in the hand, wrist or arm are strained, causing inflammation. In extreme cases, symptoms can last for years or become permanent. Writers are known to complain of the pain of writing, but some take it to extremes[1].

Arthur Munby (1828–1910) was a poet and a lawyer, but is famous today as a diarist who recorded, among other things, his fascination with working-class women. He walked the streets to converse with them, recording their stories and observing their ways, clothes and habits. He often compared

---

1 Novelist Henry James, holed up in the seaside town of Rye, in Sussex, and in the midst of working on *What Maisie Knew* (1897), a novel that tells the tale of the ugly divorce of a couple from their child's point of view, the first novel of its kind, but not the first novel to have been written using alternative means (Wilkie Collins' *The Moonstone* was both penned *and* dictated because its author was said to have been drugged-up on opiates to relieve some phantom pain). James, though, had suddenly found the discomfort caused by writing unconscionable so he took the services of a typist to whom he might dictate his novelistic reflections, and practically everyone from that day to this has noticed the shift in Henry James' style that resulted in unbelievably long sentences, loquaciously filled with information that did not always seem entirely necessary, with multiple buried sub-clauses, subjects displaced by lines from their objects, confused pronouns, embedded subordinations, verbose descriptions, and a general breathlessness throughout.

Before he took an amanuensis, James was a thoroughly modern scribe, he used to write standing up. It is unlikely that James' RSI accounts for his unique writing style, but it is a part of it – his pain is written in the genes of those long and meandering sentences that blend complex imagery with a forensic focus on the consciousness of his characters. It didn't stop him going on to dictate *The Awkward Age* (1899), *The Wings of the Dove* (1902), *The Ambassadors* (1903) and *The Golden Bowl* (1904). Henry James' RSI contributed to a late-flowering of formal adventure and productivity and proved to be loquacious in the extreme, but there is astonishing silence amongst the workers of the 19th century.

his own plump, soft white hands with those of his subjects. He noted in his journal on 21st August 1860 how a subject's right hand appeared to be "a large red lump, upon her light-coloured frock; it was very broad and square & thick – as large and strong as a sixfoot bricklayer... The skin was rough to the touch" whereas his own appeared "quite white and small by the side of hers."^

Like our strained and jangling wrists, their nerves twanging in pain, that was a working hand. Munby's pale and pudgy digits were as much a sign of class as the stooping backs and the grey faces of the factory workers.

There is also the case of the novelist George Eliot, whose odd hands have been the subject of biographical debate for generations. She was said to have possessed a right hand that was large and robust, supposedly a result of having spent much of her youth churning butter. It is not important whether or not she had such a hand (of course she didn't), but why should the question matter? Perhaps because so many Victorian codes of dress meant that apart from the head and hands, the body usually remained covered, especially women's bodies. Hands also seem as indicative of a life as the shape of a body or the lines on a haggard face, but hands in particular articulate the marks of a working life with greater eloquence than physiognomy can. After all, the hands do all the work.

While it is inevitable that the bone density and muscle mass in a favoured hand will be greater than in the other, the difference will not be noticeable; people do not grow absurdly large hands because they have churned some butter or, as in the case of the gentleman observed by Sherlock Holmes, wielded a carpenter's hammer. Championship tennis players have spent their entire career favouring one arm and one hand to stratospheric excess, but after a gladiatorial Centre Court final when Andy Murray goes to shake the hand of, say, the Princess Royal, she doesn't lurch back in horror at a proffered arm that looks like a paella pan wearing a sweatband. That's because although the forearm of most champion tennis players *is* noticeably larger on one side than the other through training, the hand is not. It's not possible to enlarge a hand in the same way as the muscles of the forearm. Despite 30 years of daily training, it is a dense, muscular, but normal-sized hand.

Now so many of us suffer from RSIs a question emerges: is it because our hands are weaker than those of previous generations? Bone density scans suggest this is the case, but because RSIs affect the soft tissues of the hand, wrist and arm, they leave practically no trace in the fossil record. We do know that early humans engaged in many repetitive and high-load activities, more so than modern humans, such as scraping animals' hides, as well as making and using stone tools, but the strength of their intrinsic muscles was certainly greater, too. We know this from the bone scans of pre- and post-agricultural women, both of whom had bones much stronger than Olympic rowers today.

What about manual labourers of the 19th century? If William Dodd is to be believed, a Victorian child labourer was less likely to have any fingers at all, "young children are allowed to clean the machinery, actually while it is in motion; and consequently, the fingers, hands and arms, are frequently destroyed in a moment."^

From 1839–40 Manchester Royal Infirmary, one of the few places that kept data, recorded 57 amputations of feet and hands in that region alone. Friedrich Engels in *The Condition of the Working Class* also recounts some horrid accidents in which children either lost a hand or died from the resulting complications.

We know that hands were so inherent a part of factory work during the period that it became the collective name for the workers themselves. But this applied equally to pieceworkers, too. While lace workers and matchbox makers, for example, were less likely to lose their hands in their work, they still worked with their hands all the time, and nowhere can I find reference to anything resembling an RSI. Even as the cascade of Factory Acts were passed during the period, it was more likely that new activities had to be found for the idle hands.

Despite constant use, our hands seem weaker than those of previous generations. While there are no bone density scans for the working Victorians, there is little doubt that the range of manual work that was normal for the majority of the population would have meant that they had more muscle and bone than we have today. Instead we work at computers all day, then at night we game or double-screen (watch TV while fiddling with a phone or a laptop).

Our hands seem as restless as our attention. It's as though we have become addicted to the limitless dopamine rewards from satisfying curiosity and we don't know how to let them go. Touching, clicking, swiping and digital manipulation are all an inherent part of this reward system (unlike other parts of our body – even our genitals), the links between the hand and the brain are disproportionately strong.

One of the profound changes we are making to the world in the meantime, (which seems to be driving most technological innovation at the moment) is that we seem to be committing enormous resources to creating an environment in which our hands will no longer be necessary. Siris and Alexas are muscling in on day-to-day tasks such as the creation of to-do lists and reminders, diary entries, sending messages, looking up movie times, adding items to our shopping deliveries; while at the other end of the scale, the dream of the technological future that is fast becoming reality is the driverless car.

For now, hands remain key in interacting with data. With phones and tablets it became possible to touch the software itself, rather than using a mouse or keyboard. And while Apple and Microsoft are both busy (as they have been for years) filing patents for hand gestures, the long game is to get rid of another link in the short chain between the brain and the processor. Facial recognition on newer phones means that our fingerprint will soon be surplus to requirements in verifying our identity. Speech recognition is not far behind.

An update to Windows 10 will add increased support for eye-tracking technology which will allow those with the right hardware to control a mouse, type using an on-screen keyboard and perform simple commands such as playing and pausing video. The software is in its infancy, but it will no doubt gather pace quickly. Digital input will soon become only a secondary mode of interacting with an artificial consciousness.

What all this means is that our hands are in the process of becoming orphaned. And what then? The one thing that modern life has taught us is that our hands do not like to have nothing to do.

In January 2018 I looked about me on a train coming out of central London and noticed that there was no one in the busy carriage not using their phone. "Look at all these phone zombies!", I thought. I threw my disdainful look about the carriage. At least, I did this for the nanosecond it took me to realize that the only reason I had looked up was because I too had been looking at my phone. Why has this become the norm in the last five years?

It is odd that phones are not ergonomic; they are designed for our pockets and not our hands. Surely our hands are changing as a result of this change in workload?

Today, articles clamour for our attention, claiming the discovery of a new mismatch disease such as "text talon" or "phone claw", but such prophesies have a long history. During the 1980s there was similar concern about "Nintendo thumb" and around the millennium there was "Blackberry thumb"; but the names never stick. And the reason they don't is that no distinct pathology emerges. People are referred to clinics and for surgery for phone-related pain, but their numbers are thankfully small. Addicted Instagrammers and obsessive Tweeters may suffer from scrolling thumb, but there are only a small minority with very serious habits.

In the 1980s and 1990s, how many people used Nintendos and other consoles too much? I'd imagine quite a few, perhaps 1–2 percent of the population. But compare this with phone use for which the numbers are huge. Yet where is the epidemic?

Because phones aren't ergonomic, there is no right way to hold one. It's just like the kinds of writer's cramp that strikes exam candidates: once your hand hurts, it's time to rest it for a bit. More than anything, RSI of any kind – be it from knitting, sewing, working a loom or sharpening a flint – means your hand needs a break. These symptoms arise from doing too much of a specific activity. The rise in cases of RSI is really connected to the way our working patterns have changed in the last 20 years.

Many of us are now mobile workers, who use time and opportunity to work a bit more on whatever is at the top of the priority list. The latest figures published by the Health and Safety Executive suggest that RSI affects 0.0073 percent of people, up from 0.006 percent five years ago; these are people who have sought treatment. The National Institute for Occupational Safety and Health in the US estimates disease rates of 7 percent, and according to them RSI accounts for about 50 percent of work-based injuries.^

A decade ago the picture looked a little different. The Chartered Society of Physiotherapy in the UK estimates its cost to employers at £300 million per year, and that 3.5 million working days were lost to the condition in 2006–7.^

These numbers may seem acceptable (£300 million is a drop in the ocean compared with the cost of back pain or type 2 diabetes, for example – or the US GDP of $1.2 trillion), but consider the research, money and resources poured into this problem by physicians, employers and health services over the last couple of decades and the idea of the numbers rising is concerning. It feels like we are breaching capacity. The environment around our hands is changing and we are reaching the point where we all perform the same tasks with computers and phones in both our work and our leisure.

The fact that RSIs cost more than $20 billion a year in workers' compensation in the US (that they are so expensive to employers and insurers) is also perhaps a key reason why we don't hear similar complaints from Victorian pieceworkers or factory hands in tales of working life. They were either too busy (and grateful) not having their hands ripped off by machinery, or spending all that they had earned yesterday on today's subsistence to worry about a bit of pain in their wrists. Neither group had recourse through employers or their underwriters.

For those who suffer from them, RSIs are debilitating and disruptive, chronic and distressing, but we are not all at risk. Nor are our hands changing or evolving as a result of using them in this way. Why? Unless the ability to use a phone becomes somehow favourable or necessary for reproduction, they are unlikely to be a priority for genetic variation.

To find out more about how our hands are an inherent part of our identity as a species, we need to look a little deeper in time.

Psychoanalyst and author Darian Leader has suggested that the reason we cannot leave our phones alone, even fidgeting with them in meetings, the cinema or theatre, is that touching them is a gesture of loss. We are rehearsing and reliving those tentative and searching gestures of our infant selves who sought the reciprocal touch of their mother. While phone use is many things, I am inclined to think that there's something in this, and I rather like the notion that through this technology we find touch and connection of different but nonetheless reassuring kinds.

When our fingers tap our screens, they are not scouring Google, swiping right on Tinder, or cataloguing their photo library, they are unknowingly conducting an eager search; they are trying to find their way home.

MEAN & PRIMITIVE

How far back do we have to go to find a hand that was substantially different from our own, so we might see how our hands have changed?

I went to meet one of the world's leading experts on the subject of hands. Tracy Kivell is Professor of Biological Anthropology at the University of Kent where she specializes in primate locomotion and skeletal morphology, as well as the origin and evolution of human bipedalism and hand use. She is leading a project which investigates the hands of living and extinct primates in the hope of discovering how some early hominin species used their hands, and consequently, how they lived. Because our hands are such a key part of our evolutionary story, this research aims to answer fundamental questions about our evolutionary history and our identity.

Using biomechanics, morphometrics and 3D imaging (technologies that assess size, shape and internal structure) Kivell and her team are trying to assess the relationships between form and function in our and other primates' hands. Their base starting point is to look at modern human hands and how they respond to external stimuli and loading (such as hanging from bars and climbing). They are also looking at the form and function of our closest DNA cousin in terms of joints (along with chimps), the bonobo.

Previous research has shown that the insides of bones have a story to tell, as they vary greatly across species which experience different environmental and evolutionary pressures on their movement and morphology. The anatomy of the hands is interesting in itself, but we also need to look inside. The data the team is looking for is a detailed map of the internal bone underneath our joint surfaces, called trabecular or cancellous bone. Bone is remodelled throughout life in response to load and use, so an accurate map of the trabecular bone can help inform Kivell's team how that bone may have been used – for locomotion, suspension or climbing, for tool use – or whether, like the modern human hand, it did not do much at all.

The same problem that meant that very few fossil feet were preserved (the bones are small, easily damaged or dispersed and tasty to scavengers) applies equally to hands. There are several relatively complete hands of modern humans and Neanderthals in the fossil record, but before that they are exceedingly rare. Kivell has an incomplete *Homo habilis* hand ("a couple of wrist bones and some fingers but that's about it"). The remainder of the deep fossil records are made up of isolated bones and "fragmentary bits from some other sites but none that we know go together in the same individual". She tells me that there is an *Ardipithecus* hand from about 4.4 million years ago which is very ape-like. Some say the claim that it is hominin is controversial because it has a divergent toe, one more compatible with grasping than bipedal walking.

Between *Ardipithecus* and modern humans there is also *Australopithecus sediba*, which was discovered by the nine-year-old son of paleoanthropologist, Lee Berger, back in 2008. The clavicle that Berger's son found turned out to belong to a two-million-year-old juvenile who stood at about 127cm (4ft 2in). This the oldest hominin hand discovered in a virtually complete form. It is in miraculous condition compared with other fossil hands; it is surprisingly modern and its morphology suggests that it is anatomically capable of a precision grip. It had short fingers similar to those of modern humans, but a much longer thumb.

Precision grip is very important for manual dexterity. This is the kind of

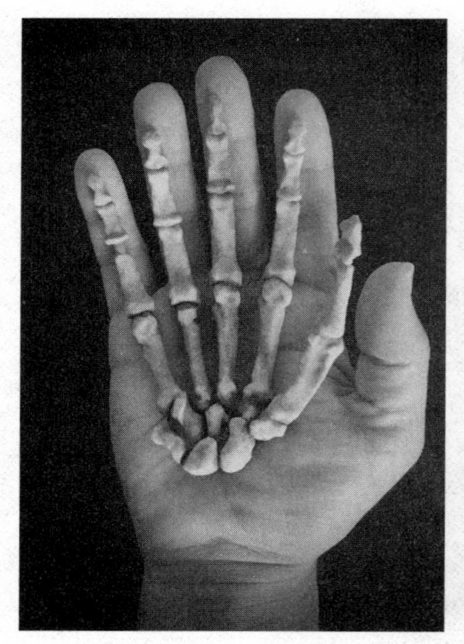
Naledi hand

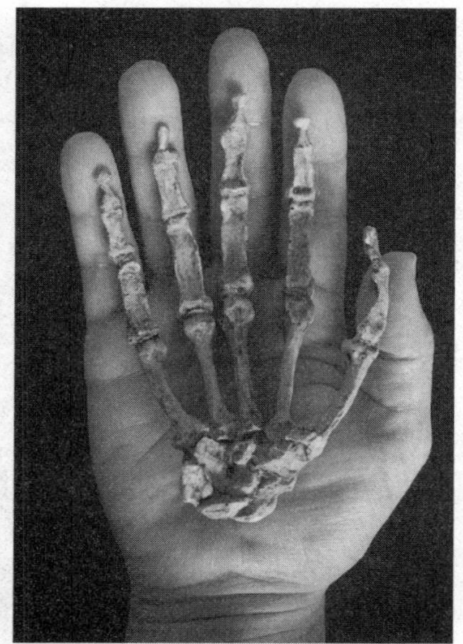
Sediba hand

grip you need to pass a piece of paper to someone without creasing it. In humans, the grip is precise and strong, and it is easy for us to modulate the power of it. It is a pad-to-pad action (thumb to finger) that other primates, such as gorillas, for example, cannot make. Instead, other primates use a tip-to-tip grip, but even with a dextrous and experienced hand, you can't, for example, write like that.

A few years after the discovery of *sediba*, Berger's team again struck gold when they found a new species: *Homo naledi* (*see* page 35). It was a confusing find. The feet were practically identical to those of modern humans, but the skulls (and so the brains) were small – about the same size as a chimpanzee's. The fossils were not immediately radio carbon-dated (the process damages the fossils), so estimates of their age were flung about with abandon. Morphology suggested anything from millions of years ago to a few hundred thousand. But they turned out to be as recent as about 250,000 years old.

Among the bones discovered was a complete hand. The thumb was long (like ours), but the fingers were also very curved, even more so than those of *Australopithecus sediba* and some other *Australopithecus* species. This is a species with a *sapiens* running foot and an ape's curved fingers.

Kivell explains that there is a possibility that an evolutionary pressure expressed through a fully modern bipedal foot meant that there might have remained a need to retain climbing ability in the upper limb if aspects of *naledi*'s life were still arboreal.

When Kivell and her team have finished mapping the bones of the hand, we should have some answers as to whether *naledi* was arboreal or if *sediba* used tools.

Variation in loads or force experienced by different parts of the hand may leave a trace in the internal structure of the bone by adding or removing tissue, but what about dexterity? The dexterity of modern humans is exceptional; from the Lion Man onward we have shown ourselves to be a species which will work to acquire and master dextrous skill and create sculpture, art and jewellery as well as learning to type at 80 words per minute or finger a Bach cello sonata.

This led me to ask the question to what extent this was unique to modern humans? Kivell explained that dexterity is common in primates when compared to other mammals. Improved dexterity likely evolved to allow them to "access certain types of food that other mammals could not. It ends up being really advantageous." So if that dexterity was already there in primates, the evolutionary advantage of becoming bipedal was huge and it makes perfect sense that it was selected for so many times in the last several million years.

"Art" then, is cognitive. The kinds of dexterity we have and that we may admire is visible in early humans. It would have taken as much dexterity to make some early tools as to carve an object or a fetish. But to endow that

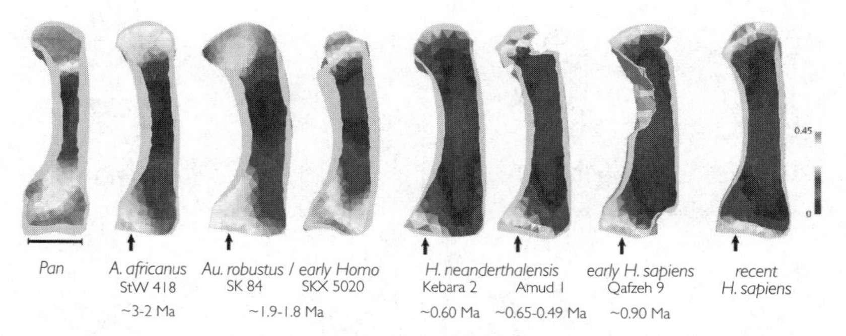

Comparison of the bone density in the thumbs of chimpanzees, early *Homo*, Neanderthals, as well as early and more recent *Homo sapiens* (darkest is thinnest) - from "Human-like hand use in Australopithecus africanus" by Matthew Skinner, *et al.*, in *Human Evolution*, 2015, 347: 6220, pp. 395-399

object with a meaning beyond its material context (a piece of mammoth ivory, for example) requires a cognitive leap; the dexterity has to be there for the leap to have somewhere to land.

I asked Kivell how our hands have changed across our species. Her answer surprised me. Almost every part of our body has responded to our environment in particular ways, but the hands look much as they did hundreds of thousands of years ago. In the matter of a generation, though, one major change has been the absence of a writer's callus (and the consequent decrease in bone density) on the second finger of the writing hand.

Modern hands might be ailing a little more than they used to; muscle mass will have atrophied and bone density decreased for most of us, but Kivell believes that there is a base level beyond which the bones will no longer decrease in density. "We used to think of human hands as being quite remarkable, but we now have a much greater appreciation for how capable apes' hands are." By implication, this means our hands are not unique at all.

She agrees. "They're actually quite primitive."

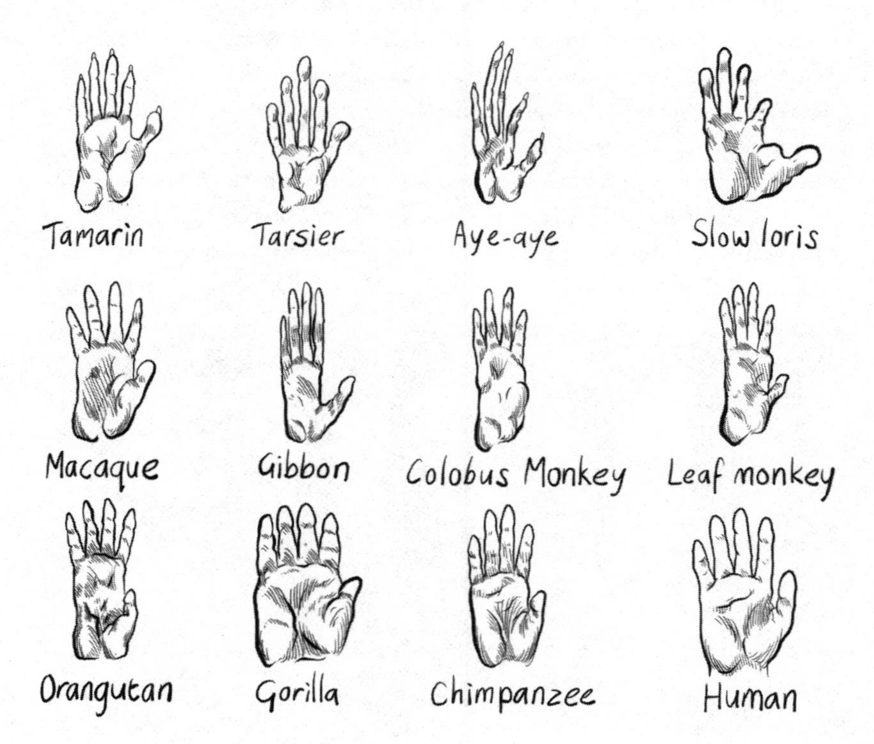

Kivell directed me to some anatomical research that had been done by colleagues who had looked at form and function in other primates. I found the ways it speaks to our story as a species simply breathtaking.

What would you get if you took the hand of a tree shrew, a lemur, an aye-aye, a galago, a slow loris, a tarsier, a tamarin, an owl monkey, a capuchin monkey, a spider monkey, a guenon, a macaque, a baboon, a colobus monkey, a leaf monkey, a gibbon, an orangutan, a gorilla and a chimpanzee then drew them all, scanned the drawings and produced an average of all the shapes and sizes? The output would be the mean and primitive primate hand, one not adapted to any particular function and probably not much good at any specific job.

This piece of research by Pierre Lemelin and Daniel Schmitt^ was written in praise of earlier work, particularly by a scientist at the end of the 19th century, Frederic Wood Jones. Jones had been the first to identify the primitive nature of the human hand in his 1916 book *Arboreal Man*.^ More than 60 years later, John Russell Napier published *Hands*,^ in which he argued:

> Man's hand shows an extraordinary degree of primitiveness, an astounding conclusion when one thinks of its specialized movements, its acute sensitivity, its precision, subtlety and expressiveness ... There is an explanation of this apparent paradox between specialized and primitive. The hand itself is derived from yeoman stock but the factor that places it among the nobles is, as it were, its connections – its connections with the higher centres of the brain.

Both men's ideas were revolutionary. They were all-out assaults on the innate vanity that wishes to believe that our hands are the finest of all mammals'. Compare them, though, to the hands of the aye-aye, with its fingers of vastly discordant sizes, the longest matching the length of its forearm. What it can do with this adaptation is terrifically impressive.

The hands allow the aye-aye to live in the canopy of the rainforests of Madagascar, but what makes them peculiarly impressive is the long thin digit. This taps the branches of trees, making a sound that allows the aye-aye to create a mental map of the hollows beneath the surface where it is most likely to find prey. The aye-aye cups its ears to pick up every tiny sound, hoping to detect the movement of a grub. Its teeth then rip away at the bark until that elongated digit can be used again as a scraper which opens up the perimeter around the food source. The long finger then works its way into the space and becomes a hook to pull the food out. It is a hand that has adapted specifically to make the aye-aye successful in their very specific environment.

Ours hands are average. There is nothing specifically adapted about them at all. They make us, as Kivell claims "jacks of all trades". The hands seem the least adapted to their environment compared with those of practically any primate and neither have they changed much throughout our time on Earth. The hands that text and type are the same as those that worked the spinning Jenny, that thumbed seeds into the soil, that carved The Lion Man, that buried the injured man at Ohalo II, that knapped the very earliest stone tools. These are your hands you hold this book in.

But here's what is different. The hands of primates have successfully evolved to take advantage of their environment. What happened to human hands is much more revolutionary.

We are unique among species because instead of waiting for evolution to provide us with the right tools for the job, our cognition reversed the process. Other apes' hands have adapted to their environment, but humans used their primitive hands to make an environment in which those hands, fingers and thumbs became the ideal apparatus for interacting with it.

Our hands have optimized their environment so that they are now the perfect appendages with which to negotiate it. The reason our hands have not changed is because there is no evolutionary pressure for them to do so, existing as they do in a time and place where they are, for now, perfect.

As technology pushes forward, and seeks ways in which our lives may be increasingly automated, a question hangs in the air: what will we do with our hands? In my lifetime I have experienced the world both without the internet and with it; is there a third phase coming in which our hands will be downgraded in their utility? In an absurd way, this seems both comically unlikely and inevitable.

We evolved. We shaped a world to be accessible to our hands, designing it specifically for that purpose. Now our hands are our key input devices for interacting with another world. As they type, they programme, they automate, and with each innovation they are gently tapping on the glass of the transhuman revolution.

What will we do as our handiwork becomes increasingly less necessary? Technology today is at a point at which the portal to the transhuman universe has already been prised open, and through software, our hands are reaching into it. They are feeling their way before they grasp and wrest the fabric, making it wide enough for us to sidle through, into who knows where...

EPILOGUE

For now, human beings, just like their hands, will continue to work because in the current economic environment they cost less than more complex machines. The robots are on the march, though – and have been throughout the two centuries since the spinning Jenny first made workers redundant, giving rise to the machine-breaking Luddites.

Today, our time is limited. The squeeze will come first from the most automatable and profitable sectors, and the change has already begun. For years, supermarket checkouts have steadily evolved toward automation. Banks are now barely recognizable – it's at least five years since I spoke to a teller. Our university library looks more like an airport security system and there are certainly no books being stamped by anyone. Transport more widely will change beyond recognition.[1]

The second Industrial Revolution has already begun. Hostility to automation and the loss of manufacturing, and related anxieties about job security in the emerging economies is making itself heard through political surprises such as the success of Brexit and Trump which few saw coming – and this is only the beginning.

In terms of our bodies, technology may help us. While it is the cause of many problems, it is often the solution to them, too. For example, screen time and page time have been associated for a long while with the epidemic of short-sightedness we are experiencing, particularly among the young. But what looked like a driving cause turned out to be only an association. Lack of exposure to sunlight is an inevitable by-product of screen and page time, because that time has traditionally been spent indoors. Screen technology, though, is quickly catching up with the real world. Phones, iPads and laptops can now reach 600 nits of brightness and more, and are resistant to water,

---

1 *Back to the Future* and *Blade Runner* both predicted that our skies would be busy with flying cars. What the films got really wrong was that people were driving them. During our lifetime the steering wheel will become ever more insignificant as our own cars, and taxis, will become machine-driven within the next few years.

making these screens easier to use outdoors. This is not a solution, but it will help to remove a few million of those framing devices from Anthropocene faces.

Wearable technology such as Garmins, Fitbits and Apple watches also seem a good solution, but have similar issues. They are great at tracking calorific expenditure, and the financial health of the industry attests at least that we are showing greater interest in fitness and at best they are succeeding in motivating many of us to move a bit more than we normally would.[2] This technology is likely to improve steeply over the coming years, but at the moment it highlights a persistent problem in how we understand activity and sedentariness.

One of the problems is that we are more likely to be attentive to our activity than our inactivity. Fitness trackers are very good at telling us how many calories we have burned, but not identifying those that we could have burned but haven't. They are great at quantifying things that happen rather than encouraging us to do those things that haven't happened because of modern life.

They also contribute to the idea that more is better: the more calories burned, the fitter you are likely to be and the more health benefits you will garner. But the relationship between fitness and longevity is far from clear, and as a higher metabolism is associated with a shorter lifespan more nuance needs to be added, not just to the technology, but to our understanding of it.[3] Perhaps the solution is to have wearable trackers that vibrate when we get on a bus, stand still on an escalator, put plates in a dishwasher or wait in for our shopping to be delivered, in the hope that our trackers will remind us of missed opportunities to move.

As movers we are slowly grinding to a halt. Automation in the house: everything from coffee machines, smart lights, TV remotes, robot vacuums, to

---

2 The first substantial study that sought to assess the role that pedometers have in motivating more active behaviour was undertaken by a team at Stanford University in the US. After a substantial review of existing research, they found that the introduction of a pedometer was found to increase the wearer's activity by approximately 2,100 steps per day. (See, Bravata, *et al.* "Using Pedometers to Increase Physical Activity and Improve Health: a Systematic Review" in *Jama* Nov 21, 2007.)

3 The "rate-of-living" theory has been around for a long time. First introduced in 1908 by physiologist Max Rubner, it sought to explain the relative lifespans of mice and elephants – for example investigating why larger animals with slower metabolisms lived longer. The ideal example to imagine is the pace of life of a giant tortoise relative to its lifespan, which is often over 150 years in length. But it is a very complicated science with a head-spinning number of factors involved. More recent research suggests that those with a higher metabolism may produce relatively fewer destructive free radicals.

washing machines and dishwashers, means that there are fewer opportunities to move inside the home. Fewer miles are travelled to and from work using foot or pedal power.

Occupational movement is also becoming increasingly rare throughout the population. A study that recently appeared in the *British Journal of Sports Medicine* found that increasing the duration of light-intensity activity could have a huge impact on "cardiometabolic health and reducing mortality risk."^The authors found that "frequent short bouts of light activity improve glycaemic control" and that "doubling light activity reduced the likelihood of premature death by as much as 30 percent." This means that if you were lightly active for 30 minutes a day and increased that to 60 minutes, your chance of premature death decreases by 30 percent. It is a diminishing-returns model; doubling one hour to two, or two to four, each produces the same effect as a 30 percent dip which becomes incrementally smaller in real terms with each step.

What is the future for the human body?

Almost every aspect of our lives is saturated with technology. The basics – the way we move, consume, communicate and seek out information – have all changed dramatically in response to technology, whether we look at the last few centuries or even just the last few years. And while it is tempting to believe that if only technology advanced enough, it would solve our biggest problems, some philosophers have suggested that the belief in technology lies at the root of the problem.

THE QUESTION OF TECHNOLOGY & DICKENS'S HAUNTED MAN

Martin Heidegger (1889–1976) was one of the most important philosophers writing about modernity and modern life, who expended a good deal of energy in trying to understand our relationship with technology, the world it created and the humans we became as a result. It was not things or stuff, and our attachment to them that he was interested in, but in how technological thought and belief had become an inherent part of being human.

Heidegger saw technology as the constriction of experience. He felt it was something that got between humans and the world so that when they looked at it, the idea of using the world, rather than being in it, predominated.

Everything: nature, the environment, even other people are seen as technological phenomena. And the cloud of technology spreads widely across all our thoughts and ideas, into the future of course; but it also shapes the past in that our way of relating to it is informed by present concerns. (The same might also be said of religious belief.)

It was not all technology that Heidegger was interested in, but that of modernity. He did not see early tool use by hominins and the development of craft and trading skills as inherently problematic; instead he was concerned with how the Industrial Revolution changed the way we related to the world around us.

Heidegger believed that after the Industrial Revolution the world turned into a kind of standing reserve of energy, and instead of being in it, humans were simply waiting to acquire its use in some way, at which point everything was requisitioned into the project of modernity. This idea encompassed many categories, ranging from resources such as fossil fuels to places as disparate and seemingly innocent as the "wild" and "natural" spaces that humans need as places of recreation or to maintain a given species in the Anthropocene.

Wild places, in Heideggerean terms, are only those permitted or designated to remain wild because it is useful to our species in some way for them to remain so. Usefulness dominates absolutely our creative, intellectual and economic activity: technology is a way of being, and nature is the storehouse of energy devoted entirely to the project of modernity.

For Heidegger, the realm of experience that preceded technology had gone, but we might find glowing embers of it in the activities practised by our earliest ancestors: in art, poetry, simple sculpture, ritual – and in experiencing, accepting and revelling in the limits and pleasures of the human body. These were all routes to "unconcealing" the world about us. Our bodies are the key to this unconcealing and yet it seems that modernity, not content with changing the world, has changed us, too. Our hands have been at the very forefront of creating this change.

Life moves fast; evolution is slow. And while there have been some evolutionary changes to our bodies in the last few thousand years, they are insignificant in terms of adapting to the needs of modern life. By the time random mutation gets round to sorting out our backs or thickening our tooth enamel, we are likely to have changed our diet and working practices, will not be around any more, or will have become transhuman.

One development has been that our work places and practices are increasingly selecting certain kinds of mental attributes that align with success in specific fields. There have been studies in Silicon Valley of the usefulness and effectiveness of certain kinds of autism and attentional variation which seem particularly productive in coding circles. Particular body types (among them linear ectomorphs, rounder endomorphs and more muscular mesomorphs) and their related pathologies will increasingly find expression in sexual selection, with endomorphs inevitably struggling in an increasingly sedentary future.

In a world in which it is likely that genetic code will soon become as editable as HTML (this is already being trialled in human patients), what will we do about anxiety, depression and low pain thresholds? These were all incredibly useful to the Holocene and Pleistocene human, especially those who needed to be wary of predators. Are they much use to us now? Do we need to sweat or get short of breath as a result of a job interview, or worry because we are a little late?

We must be cautious in how we proceed. Geneticist Adam Rutherford explained succinctly how far we had gone with gene therapies in 2015: "the number of diseases that have been eradicated as a result of our knowing the genome? Zero. The number of diseases that have been cured as a result of gene therapy? Zero."^ But the time is coming, and soon.

There is a wonderful, little-known Christmas book by Charles Dickens called *The Haunted Man*.^ The protagonist is a man of science called Redlaw who is saddened by the memories of those lost to him and wishes never to be burdened again by unhappy memories. His wish is granted. His sad thoughts drift away, but his newfound contentment has some unforeseen consequences. He is no longer able to relate to those around him. His skills of empathy vanish; he is cruel and cold. Without the memories of the hard times, those losses and failures, Redlaw is unable to connect to other people, is often inexplicably angry and as he passes his "gift" on to those around him he witnesses their terrible behaviour, too.

Balance is restored in the novel when the ghost turns back the clock allowing everyone to remember all their unhappy experiences as well as the happy ones. Peace reigns when they recognize that "the memory of every remediable sorrow, wrong, and trouble in the world around us, should be active with us, not less than our own experiences, for all good."

The story is interesting because it is an analogue version of a digital transhuman hack. It is as if Redlaw has gone into his DNA, deselected the parts he feels are the cause of the trouble, and has unknowingly wreaked havoc in a system he misunderstood. The story suggests that we need unpleasant physical and emotional states, because without them (as Redlaw learns) empathy, curiosity and fascination are lost.

It's a simple tale that warns of an obvious truth: our bodies are more sensitive, in the truest sense of the word, to change than we might think. The knowledge that we have acquired over a few thousand years seems vast. With each generation that knowledge is built on and passed down to the next. But can it ever compare to the complex ingenuity our bodies have developed over many hundreds of millions of years of evolutionary experimentation and adaptation? Because of their complexity, our bodies will persist in being less editable than we might believe them to be and this leaves us with the important question of what we can do about them now.

We are only at the very beginning of understanding the plasticity of epigenetics, and with every month that goes by we learn more and more about how the genes that we mistakenly believed were a roll of the dice at the point of conception are more like meerkats waiting for a signal in their near surroundings to respond to. The key mode of operation in epigenetics is methylation, which activates or deactivates parts of a gene, but numerous aspects of its heritability are yet to be discovered. Although the DNA remains unchanged, the methylations modify its behaviour. Science is already showing how the things done to a body in one life can be passed down to the next.

The Anthropocene human burns about 200–300 calories per day above their basic metabolic minimum. During the Paleolithic this figure was probably about 1,000 and for Neanderthal males, as much as 2,000. When the Anthropocene human burns the few calories it does, it might be on a walk to the station, or some minor chore or activity which punctuates long periods of inactivity that is not dissimilar to mammalian hibernation. The Paleolithic human burned calories at a relatively low level of exertion throughout the day.

After all we have learned, perhaps we might consider renaming our species? ***Homo sapiens ineptus***: human, smart, but not really compatible with the abundance of knowledge, food and comfort in our environment.

WHAT IS TO BECOME OF US?

Meet Joe. The unusual thing about Joe is that "Joe" is not his first name, but his surname. His friends actually call him Average. Average Joe

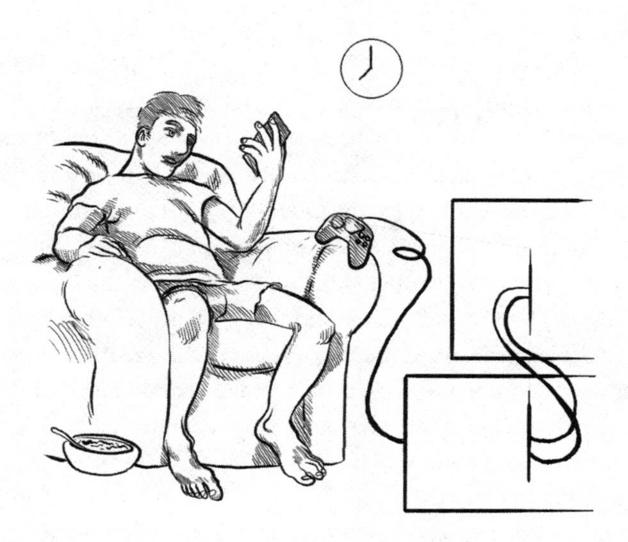

is middle-aged, has a normal job for which he leaves the house at eight in the morning and drives to work, which takes about an hour. Once there, he works at his desk until it's time to eat. Then he goes to the cafeteria in his office block for lunch. He returns to his desk and works until it's time to start the drive back home, but often stays late as he does not want the boss to leave before him. On his way home, he stops off at the gym to do a little weight training for half an hour, then goes home for dinner. By now he's quite tired so spends a restful evening gaming, reading and watching TV before retiring to bed.

Joe goes to the gym nearly every day, and on the days that he doesn't go, he likes to take a walk because he thinks that activity is good for him. Joe believes he is fitter than most of his colleagues at work because they seem to do no exercise at all.

Joe doesn't realize it but he is sitting for approximately 15 hours a day as he stacks up the hours at work and then rests when he gets home. According to research conducted by the British Heart Foundation, the behaviour of Average Joe is normal for adults employed in sedentary work.^

We now know that the likely outcomes for Average Joe are: some pathology in late middle age – back pain is practically a certainty. When he does retire, he is more likely to need the services of a full-time carer or nursing home, but that presumes he lasts that long. Research from numerous sources, groups – even countries, shows that adults who sit for substantial parts of their day are at a higher risk of early death, particularly from cardiovascular disease.

To avoid this future, all Joe has to do is to persist with his exercise regime while introducing further movement throughout his day. He will have to either work or game a little less, because his body needs more active time than he's currently giving it. He needs to interrupt his sitting time frequently and regularly while at work. If Joe does these easy and simple things, they will change the latter third of his life substantially.

If Joe decides not to bother, and wants to show his boss how very committed he is by putting in more hours at the office, then he will likely find himself being prescribed medication in middle age that he will stay on in perpetuity, or until that medication is replaced and upgraded with more potent alternatives. One of Average Joe's friends should tell him that he works too hard; that it's not good for him.

The research on inactivity that I read while writing this book had a profound impact on me. As someone who works a sedentary job I realized I would have to work out ways to move more. It took a few weeks to get into the routine, but I found a way of fitting more activity into my day.

I have tried to shake the idea that the best place to work is at home. Now, on an average day I will write in at least three different places; one of them is always home, but the others are usually spread out over a few miles and I walk between them. So, I now walk 64–80km (40–50 miles) a week and it feels like nothing. It's easy; I don't feel tired; it has kept me sane and I'm convinced it has played a significant role in reducing my back pain.

My back is not "cured", but the bouts of pain I have now are more rare and short-lived, and I am convinced that the walking is the reason. The idea that pain might be created by something over which we have a modicum of control makes a huge difference to the anxiety that automatically adheres to the stress we feel when experiencing any kind of pain. While the back pain had once been chronic, I can often get through most of a day without thinking about it. More significantly, months have gone by since I worried about it. I still go to the gym and run, but with less vehemence as they don't seem as necessary – both are recreational rather than functional.

I don't want to be Average Joe any more, whose future may be counted in tottering reams of prescriptions, like so many others.

Currently, more than half the adults in the UK take prescription medication, and in the US that figure balloons to 70 percent (about 210 million). Laid end to end, that's enough pills to orbit the planet twice every year. In the UK, the majority of drugs prescribed are statins, beta-blockers and painkillers. More worrying are the alpine mounds of antidepressants prescribed, with some worrying patterns emerging in the way they are distributed: as many as one in five women on low incomes take them, for example. My asthma medication appears as number 32 in the world rankings, while Viagra stands proud and firm at a surprising 40.^

At the very least, these figures tell us that we are too eager to treat when we should commit greater effort to creating and inhabiting a more preventative environment. One of the most pressing questions for governments around the world is to consider to what extent they wish to invest in prevention, or whether they would rather continue to shovel piles of cash into the furnace of morbidity.

Billions is being thrown at genetic research and gene therapies. This is the future and it is worthwhile research, but it makes much more economic and holistic sense to think about exactly how so many diseases are emerging from the fog. We are just cresting the moment when genes are becoming editable. And the history of science teaches us that with many interventions, such as the C-section, come related and unanticipated consequences and associated problems that often do not emerge for decades or generations (such as allergies).

For all that is understood about genes, their presence is not a guarantee of an outcome in the phenotype (the physical manifestation of genetic code). There is no single gene for brown eyes and there is no single gene for ginger hair. The genotype of any individual might contain markers that make these outcomes likely, but not a certainty. For now, genetic editability is complicated by the fact that genes work in concert with one another in ways that are going to take time to untangle.

Alongside genetics, the other key player in the emergence of much disease is the environment. Not only is it highly malleable, but we don't need hyper-sophisticated gene-editing tools such as TALENs, ZFNs or CRISPRs to change it. In most cases, we just need to use our environment a little differently. Why risk the complex and mysterious associated pathologies that will probably emerge from gene editing when a walk to and from work might solve the problem?

The World Health Organization estimates that our annual healthcare bill amounts to about $7.2 trillion globally. The US alone spends about $8,362 per person per year on healthcare.^ I could save the governments and people of the world at least 85 percent of these costs. My silver bullet would roll back at least 90 percent of the morbid conditions that plague us. I can outline my solution in ten words. **Governments of the world: address falling activity levels and obesity.** And if we do that, nearly all the big killers will waste away and we'll have about $5.5 trillion dollars in the pot for a really great party.

In the meantime, it's down to us to do what we feel we can. For me, and no doubt for most people reading this, the problem is time. One of the most utterly foul inventions is the idea that "time is money". It is a wealth-maximization strategy that emerged at the very beginning of the Industrial Revolution.

Being wary of nostalgia for a time that was no doubt extremely tough to endure, when people lived under the feudal system, the rural and agricultural economy meant that goods and services were usually exchanged at arm's length (a handmade society). This meant that there were periods of intense work and high productivity, such as harvesting or haymaking, which were often followed by periods in which both the land and the people could recover.

A book called *The Fable of the Bees: or Private Vices, Public Benefits* by Bernard de Mandeville^, a philosopher and man of letters, caused a scandal when it was first published in 1705, offending almost everyone except Dr Johnson. It laid out what the modern capitalist economy was about to become:

> ... in a free Nation where Slaves are not allow'd of, the surest Wealth consists in a Multitude of laborious Poor, for besides that they are the never failing Nursery of Fleets and Armies... as they ought to be kept from starving, so they should receive nothing worth saving... It is the Interest of all rich Nations, that the greatest part of the poor should almost never be idle, and yet continually spend what they get... To make the Society happy and People easy under the meanest Circumstances, it is requisite that great Numbers of them should be Ignorant as well as Poor.

Those who fulminated against the book were like Caliban raging at his own reflection. *The Fable of the Bees* had the gall to articulate fully the capitalist project. A few more philosophical contributions over the decades and capitalism was soon in full sway, blowing the feudal world to matchwood in a storm of uprisings and revolutions that swept across Europe. Idleness was not only bad for business and bad for your employer, it was now bad for your country. Clocking off half an hour early on Friday was elevated from being a moderate indulgence to a kind of treason.

By the mid 19th century, work – even the meanest sort – had become a serious business. Bradshaw's Guides were a series of railway timetables and travel guides for the burgeoning tourism industry. One publication was *Bradshaw's Handbook to the Manufacturing Districts* (1854)^, which advises travellers (or as we might call them today, poverty tourists) to proceed with caution in some of these factory districts. But it was not the opprobrium of the workers that they should be wary of:

> Unless the applicant has a letter of introduction from some person known to the proprietors, there is considerable difficulty in the way of obtaining admission. As it has been justly observed by a local authority, "the objection generally entertained is not found so much upon a fear of admitting a person who might take away information that the owners wish to monopolise, as it proceeds from the fact that visitors occupy the time of an attendant, and disturb the attention of the operatives throughout the mill. The loss accruing from this cause is frequently more than can be readily estimated.

One's time certainly is money, but in most cases it is someone else who benefits. Today, in the gig economy you are unlikely to see the well-heeled taking advantage of these "new and convenient" ways to work. The freedom offered by mega-corporations (that neither own nor produce anything) to their casual employees only appeals to the poorest and most cash-strapped. The characters' names have changed, but the script seems much the same.

What we see in the Anthropocene body is, in many cases, the result of this so-called work ethic that quietly whispers in our ear to work harder and longer, and to judge the activities our bodies desperately desire us to do as that fiendish and revolutionary thing: a waste of time.

The time spent working now is killing us. In the UK, Europe and the US toward the end of the 19th century, there was agitation to enforce an eight-hour working day. Understandably, the employers were against it, as were some of the working classes at the lower end of the income scale, worried as they were that so few hours would not pay them enough to subsist on. Today "9–5" hours are synonymous with a normal working day or, indeed, working life. But the eight-hour day remains only an ideal for most. With wages stagnating, workers in the UK, the US, Australia and Europe are working longer hours for the same pay. Where there are pay rises, particularly in the public sector, they usually lag several percentage points behind inflation, and with technological advances many aspects of our jobs get harder instead of easier.

We live a tethered existence. Technology sends information across the globe in milliseconds, and like a crossbow bolt sent from the gods, it can find us wherever we are. Working long hours is bad for us. Long hours have been strongly associated with health risks such as heart disease. A study called "Are vacations good for your health?"^ correlated "frequency of annual vacations by middle-aged men at high risk for [coronary heart disease]" with a reduced risk of all-cause mortality and, more specifically, mortality attributed to [coronary heart disease]. "Vacationing", the paper concluded, "may be good for your health."

When history looks back on the weird life we all lead in offices, it will see that these were an environment as toxic as those in 19th-century factories. There might be less fluff floating in the air, but the fact that most office blocks are firmly rooted in highly polluted areas does not make them much safer. They are greenhouses for biomechanical problems, as well as for big killers such as depression and stress.

The American Institute of Stress has estimated that the condition costs the US economy $300 billion each year.^ This is little surprise when most office workers have just as little autonomy as factory hands once had. Then it concerned Ruskin and Carlyle; now it should concern all of us.

Workplace injuries are down, but workplace violence is up. The United States Department of Labor warns: "Nearly 2 million American workers report having been victims of workplace violence each year."^ Office life, at least the way that we experience it, does not agree with us. In the 19th century, there was rarely any comeback for workers injured or killed by negligent factory owners. But the legislation which began in that period has since matured and we are now comparatively well-protected. Mental health issues are just as measurable as physical ones, and there is no reason why employers and governments cannot do more to monitor the effects of stress, presenteeism and the long hours' culture on the workforce. To start tackling the problem globally, we need first to know what it is.

Everyone knows that long hours don't really work. They do not lead to greater productivity. Charles Dickens penned 14 "big" novels, countless short stories and at least five novellas. He ran several periodicals on the side, wrote a few plays, took part in amateur theatrical productions, gave public performances of his work, wrote thousands of letters, walked up to 20 miles a day, fathered ten children, knocked off a few poems and histories – and yet he didn't work in the afternoons. Indeed, it was Charles Darwin who cited that he did his best work while strolling round his garden.

Might Joe be a little less Average and more exceptional if he wasn't so tired and micro-managed all the time?

As we have seen, it would be a mistake to think of our calorie intake and the way our bodies have used them over time as remaining unchanged and static (this altered when eating times in the West changed in the 18th century and again in the 1980s). So, too, is it a mistake to think of our working patterns as unchanging and inevitable. The cures are not to be found in yoga, a standing desk, a bit of Alexander Technique or static stretching. What we really need to do is overhaul the ways that we work so they fit better into the ways we need to live.

For someone trying to understand what to do with their body to ensure its longevity, it can be a little like sitting too close to the brass section of an orchestra. The chorus of advice is deafening. We are told to perform high-intensity exercise, but in terms of longevity this is not ideal. Athletes do not live a long time, but they fare better than sedentary people. The key seems to be in the type of exercise we should aim to do for long-term good health.

The message is not getting out there about fitness. Most people do not understand what inactivity is and equate it with being unfit or not having exercised that day. But fitness and inactivity are in fact complementary states. Like Average Joe, people might be taking all the risks associated with sedentary behaviours while going to the gym five times a week. They will be better off than the person who does not go to the gym at all, but are still at risk of all of the diseases associated with sedentary behaviour. And these behaviours are encouraged by our environments.

In 2010, the *American Journal of Preventative Medicine* published a study of 5,556 people^ which found that the groups of people watching the most TV (14+ hours):

> ... were identified and described using a combination of demographic (ie lower household incomes, divorced/separated); health and mental health (ie poorer rated overall health, higher BMI, more depression); and behavioural (ie eating dinner in front of the TV, smoking, less physical activity) variables. The subgroup with the highest rates of TV viewing routinely ate dinner while watching TV and had lower incomes and poorer health. Prolonged TV viewing also was associated with perceived aspects of the neighbourhood environment (ie heavy traffic and crime, lack of neighbourhood lighting and poor scenery).

The environment is strongly linked with sedentary activities that are gateways to other pathologies. People should be free to indulge in sedentary activities – there is nothing wrong with that – but when the human-made environment seems to predetermine sedentary behaviours, that should at least warrant a second glance.

We want simple answers, such as exercise; we want to hack our lifestyles with a single line of code (as Dickens's *Haunted Man* did) in the hope that this will satisfy the needs of the tens of trillions of cells in our bodies. The quick-fix culture we live in sells us the idea that a high-intensity workout can be completed in as little as seven minutes so that we can get back to work all the quicker. And while the data shows that there are metabolic benefits of doing this rather than nothing, the idea that you can exercise for a few minutes a week to offset the effects of an otherwise sedentary lifestyle is not only ridiculous, it's irresponsible.

High-intensity interval training (HIIT) will work a kind of magic on your body which will be a whole world better than remaining sedentary. It will improve insulin sensitivity; cardiovascular and metabolic indicators also suggest it is good at streamlining our use of oxygen during exercise and increasing muscle mass, for example. These are all good things. But HIIT will not stop you from being sedentary, restore motion to joints immobilized by life or strengthen your heart as well as low-intensity endurance training can. It will not accustom muscles to doing the job they are supposed to do or provide

you with the huge benefits to mental health associated with outdoor exercise. There are so many of these benefits that hardly a day goes by without some new research that confirms the quiet and subtle ways it has of rewarding you.

The Black Dog Institute in Australia recently completed a trial involving 34,000 Norwegians who were tracked over 11 years which discovered that just one to two hours of exercise per week "can deliver significant protection against depression. "^ Bear in mind that depression is an illness which the World Health Organization believes affects 300 million of us.

A trial run by environmental scientist Mike Rogerson at the University of Essex involved working with the Wildlife Trust and tracking the mental health of volunteers who were prescribed work on the scheme to improve wellbeing. The trial subjects dug ditches and assembled bird tables. When the study began, 39 percent of the participants reported low wellbeing. After 12 weeks of getting their hands dirty, the number had halved to 19 percent.^ Our bodies are craving to spend more time in surroundings that for hundreds of thousands of years, they both recognized and felt at home in.

Instead our environment, the concrete, the office blocks, are seriously stressing us out. Inflammatory responses in the body are quite normal. Inflammation is part of the body's defence against injury or infection, and it usually involves (deliberately) sensitizing nerves so the damaged area becomes very painful, to signal to the owner that it needs a rest. Unfortunately, in some cases, the inflammatory reaction is often not dampened down at the end of the healing period. (And the bad news is that as you get older this happens more, with the result being pain that can persist for months or years.)

Inflammatory reactions, then, are part of the immune system's emergency service. When something is wrong, it is normal to find raised levels of C-reactive protein (CRP) in the blood, which provides evidence for a recent immune inflammatory response. People who suffer from allergies, who are always fighting inflammation of one kind or another, will often have elevated levels of CRP in their system. In 2003, Nader Rifai and Paul Ridker published some rather worrying research in *Clinical Chemistry* which found inflammatory responses in otherwise healthy individuals throughout the population of the United States. The people in their study did not have heart disease or chronic allergies, they were supposedly healthy yet their CRP levels indicated evidence of a sustained inflammatory response.^

This is most unusual and quite alarming.

When these findings are compared with another study which found a pattern inconsistent with the one of constant inflammation in the US they suggest that something is awry in the Anthropocene human. The findings documented "a pattern of variation over time that is distinct from prior research, with no evidence for chronic low-grade inflammation."^The region they studied was lowland Ecuador, and tracking their subjects over weeks at a time, the study found that levels of CRP rose during infection and returned to zero when the subjects returned to health.

This research might be telling us that the Anthropocene human has changed its ability to regulate its immune system, leading to a chronic state of inflammation. Inflammation is bad news; very bad news. It's associated with a much greater likelihood of depression, type 2 diabetes, cardiovascular disease – the list goes on and it's all bad.

The exact cause for the worrying disparity in these two studies is for the time being impossible to say, but it seems obvious that industrialized life is making us sick; or at the very least, it is making our bodies respond as if they are.

The 20th century – and to a lesser extent the 19th – witnessed huge advances in the understanding of the natural sciences, which led in turn to a proliferation of technologies that seeped into every aspect of human life: diet, medicine, recreation, sex, communication, everything.

We are already cyborgian – the transhuman revolution has begun. Since the launch of the iPhone on 29 June 2007, we are more connected than humans have ever been before. We have more knowledge at our fingertips than any of us can possibly know. I can ask my smart watch to make me a cup of tea; admittedly, it won't do anything, but it can listen, interpret and understand the question and unhelpfully reply in the negative.

In this hyper-connected world epidemics of loneliness are flourishing. Our hands hold a galaxy filled with people and information, always open, primed and ready for us, but instead of satiating us it is freaking us out because we feel we are always the ones standing alone on the driveway while the party rages on behind locked doors without us.

The range of technological possibilities we carry around in our pockets encourage us to live lives that are more sedentary than anything known in the 2.3 million years of our history. Smartphones, laptops, computers and the internet have not just opened up new possibilities for commerce, they have not just changed the way we do things, they are changing our bodies, our lifestyles, our longevity. It is not a leap to say that these technologies are changing who we are as a species.

We are already into the next phase of human development and if we want to take our bodies with us into this future, we must attend to them. We must try to understand not just their desires, but new and subtle ways of providing them with what they were built to enjoy, with what will help them to thrive.

Our hands have made this place that we now live in, and as much as they have built this world; they are also choking the life out of it and us, too. It's as if they are trying to kill the host.

As we try to orientate ourselves among the debris washed up by the Anthropocene, we need some grand gestures that will change the world we have made.

How can we make sense of the fact that at the end of the Pleistocene (12,000 years ago), the Earth's human population dipped perhaps as low as 10,000 but once the weather turned in favour of our species, it has since ballooned by about 730,000 percent? Estimates vary, but according to UN predictions, world population is forecast to balloon by the end of this century. Such an increase would make that population growth figure 1,120,000 percent.^

We need something as bold as the Clean Air Acts of the 1950s: trenchant, radical, upsetting, discomfiting and life-changing. At the beginning of 2018, Germany's main administrative court upheld Stuttgart and Düsseldorf's decision to ban diesel cars because of the pollution they cause, which affected 12 million motorists in the country. Germany currently has 70 regions which have breached EU limits on NO_x. It's too early to see if this is the beginning or just a bump in the road. Grand gestures like this one will help us create new habits: small changes that make a big impact. These habits might take a little time to adjust to but could add years of good life.

If a wide range of health issues, such as obesity, mental illness and premature death, are due to our man-made environment, then it's time to stop apportioning individual blame to those whose choices are narrowed and limited by their circumstances and because they exist in pathogenic and obesogenic environments.

It is time to look more closely at the world we have made, think about how we use it and reassess what we expect both from it and our bodies – and change so that we might revel in a little more of what the great forests of the Earth once readily offered us.

ENDNOTES

WHAT BECOMES YOU?

PAGE 11: "vegetated throughout a passably comfortable existence..."
Friedrich Engels, *The Condition of the Working Class in England* (London: Penguin, 2009), p. 51.

PAGE 12: ...while some studies suggest that their feet cover 1km (0.6 of a mile) *over an entire month.*
Public Health England, "6 million adults do not do a monthly brisk 10 minute walk" www.gov.uk/government/news/6-million-adults-do-not-do-a-monthly-brisk-10-minute-walk (accessed 10 Jan 2018).

PAGE 13: In *The Analysis of Beauty...*
William Hogarth, *The Analysis of Beauty: Written with a View of Fixing the Fluctuating Ideas of Taste* (London: 1753).

PAGE 14: These changes to the stress-shielding capability of the intervertebral discs...
Mike Adams and Patricia Dolan, "Spine Biomechanics", *Journal of Biomechanics*, 2005, pp. 1972–1983.

PAGE 22: I creaked my way around the classroom...
Charles Dickens, *Our Mutual Friend* (London: 1864).

PAGE 22: "little of stature, ill-featured of limbs" and "crook-backed".
Thomas More, *History of King Richard III* (London: 1557), p. 37. See, www.bl.uk/collection-items/thomas-mores-history-of-king-richard-iii (accessed 2 Feb 2018).

PAGE 23: Back pain, specifically lower back pain...
See, George E Ehrlich "Low Back Pain", *Bulletin of the World Health Organization* 2003, 81:9, pp 671-676; Health and Safety Executive, "Work-related Musculoskeletal Disorders (WRMSDs) Statistics in Great Britain 2017", www.hse.gov.uk/Statistics/causdis/musculoskeletal/msd.pdf, (accessed 14 Mar 2018); Office for National Statistics, "Sickness Absence in the Labour Market: February 2014" (accessed 14 Mar 2018); Rodrigo Dalke Meuci, *et al.* "Prevalence of Chronic Low Back Pain: Systematic Review" *RSP,* 2015, 49:1; "Back Pain 'Leading Cause of Disability,' Study Finds", March 25 2014, www.nhs.uk/news/lifestyle-and-exercise/back-pain-leading-cause-of-disability-study-finds/ (accessed 7 Mar 2018); and, Damian Hoy, *et al.* "The Global Burden of Low Back Pain: Estimates From the Global Burden of Disease 2010 Study", *Annals of the Rheumatic Diseases*, ard.bmj.com/content/early/2014/02/14/annrheumdis-2013-204428 (accessed 7 Mar 2018).

PART I

PAGE 27: The name by which our species is classified runs thus...
Human taxonomy is a contentious issue with some definitions running to above 30 categories, this is: kingdom, phylum, class, order, sub-order, infra-order, family, genus, species. For some discussions which focus on humans and primates, see, S Cachel, *Fossil Primates (Cambridge Studies in Biological and Evolutionary Anthropology)*, (Cambridge: Cambridge University Press, 2015); Jared Diamond, "A Tale of Three Chimps" in *The Third Chimpanzee: The Evolution and Future of the Human Animal* (London: HarperCollins, 1992); E Mayr, *Taxonomic Categories in Fossil Hominids*, (Cold Spring Harbor Symposia on Quantitative Biology, 1951), 15, pp. 109–18 and, Don Simborg, *What Comes After Homo Sapiens? When and How Our Species Will Evolve into Another Species* (Mill Valley: DWS Publishing, 2017), pp. 23-8.

PAGE 28: According to palaeontologist Stephen Jay Gould...
Stephen Jay Gould, *Wonderful Life: The Burgess Shale and the Nature of History* (London: Penguin, 1991).

PAGE 28: The phenomena was observed as long ago as the 19th century...
Charles Darwin, *On the Origin of Species* (London, 1865), p. 333. See, Simon Conway Morris, "The Cambrian 'explosion': Slow-fuse or megatonnage?" *PNAS*, April 25, 2000, 97:9, pp. 4426-4429; William Buckland, *The Bridgewater Treatises on the Power and Goodness of God as Manifested in the Creation* (Treatise VI), 2 vols (London, 1836).

PAGE 28: Its significance is such that a team headed by Professor Simon Conway Morris...
Simon Conway Morris, *et al.*, "Pikaia gracilens Walcott, A Stem-Group Chordate from the Middle Cambrian of British Columbia", *Biological Reviews*, 2012, 87:2, pp. 480-512.

PAGE 32: In *Extinct Humans...*
Ian Tattersall and Jeffrey H Schwartz, *Extinct Humans* (Boulder: Westview, 2000).

PAGE 35: More recently, in 2016...
Kevin G Hatala, *et al.*, "Footprints Reveal Direct Evidence of Group Behavior and Locomotion in Homo Erectus", *Scientific Reports*, July 2016, 6.

CHAPTER 1

PAGE 36: In *The Descent of Man*...
Charles Darwin, *The Descent of Man* (London, 1871).

PAGE 38: ...as Andrew Whiten argued in his essay...
Andrew Whiten, "The Evolution of Deep Social Mind in Humans" in *The Descent of Mind: Psychological Perspectives on Hominid Evolution*, eds Michael Corballis and Stephen E G Lea (Oxford: Oxford University Press, 1999).

PAGE 39: We know from studies and stories of feral children...
See, Michael Newton, *Savage Girls and Wild Boys: a History of Feral Children* (London: Faber, 2002).

PAGE 41: Research published in the *Philosophical Transactions of the Royal Society* in 2013...
Teruo Hashimoto, *et al.*, "Hand Before Foot? Cortical Somatotopy Suggests Manual Dexterity is Primitive and Evolved Independently of Bipedalism", Philosophical Transactions of the Royal Society, 19 November 2013, 368: 1630.

PAGE 43: Other research by paleoanthropologists (particularly Daniel Lieberman and Dennis Bramble)...
Dennis Bramble and Daniel Lieberman, "Endurance Running and the Evolution of Homo" *Nature*, Nov 18 2004, 432, pp. 345-52.

PAGE 49: Physical anthropologist Erik Trinkaus...
Erik Trinkaus and Hong Shang, "Anatomical evidence for the antiquity of human footwear: Tianyuan and Sunghir", *Journal of Archaeological Science*, 2008, 35;7, pp. 1928-1933.

CHAPTER 2

PAGE 53: Wolff's Law tells us...
The law of bone modelling and remodelling is one of the most widely-cited laws where so few have had access to the original. Julius Wolff's *The Law of Transformation of the Bone* was published in German 1892, with only a few copies printed. John Nutt, *Diseases and Deformities of the Foot* (New York: E B Treat, 1915), p. 157; is the earliest English language outing for it that puts it side-by-side with Davis's law, in Julius Wolff's *The Law of Bone Remodelling* was first published in English in 1986, translated by P Maquet and R Furlong (London: Springer, 1986).

PAGE 54: Here it is, used for the first time in John Nutt's 1913 study...
John Nutt, *Diseases and Deformities of the Foot* (New York: E B Treat, 1915), p. 157.

PAGE 57: In the 1960s anthropologist and ethnographer Marshall Sahlins...
See, Marshall Sahlins "Notes on the Original Affluent Society" in *Man the Hunter*, eds R B Lee and I DeVore (New York: Aldine, 1968), pp. 85-9.

PAGE 57: There has even been the suggestion that what drove human innovation was exactly this: boredom.
See, Karen Gasper and Brianna L Middlewood, "Approaching Novel Thoughts: Understanding Why Elation and Boredom Promote Associative Thought More Than Distress and Relaxation" *Journal of Experimental Social Psychology*, May 2014, Volume 52, pp. 50-57; Sandi Mann and Rebekah Cadman, "Does Being Bored Make Us More Creative?" *Creativity Research Journal*, 2014, 26:2, pp. 165-173; and Charlotte C Burn, "Bestial Boredom: A Biological Perspective on Animal Boredom and Suggestions for Its Scientific Investigation" in *Animal Behaviour*, August 2017, 130, pp. 141-151.

PAGE 57: The research was the first that scanned bone density from specimens...
Habiba Chirchir, Tracy L Kivell, *et al.*, for "Recent Origin of Low Trabecular Bone Density in Modern humans", *PNAS* January 13, 2015, 112:2, pp. 366-371.

PAGE 58: Research published in November 2017...
Alison A Macintosh, *et al.*, "Prehistoric Women's Manual Labor Exceeded that of Athletes Through the First 5500 Years of Farming in Central Europe" *Science Advances*, 29 Nov 2017, 3:11.

PAGE 60: In the Anthropocene, that figure lurches between 1,000 and 10,000...
"Species Loss: Wetland Species Disappear" *World Wildlife Fund*, wwf.panda.org/our_work/water/freshwater_problems/species_loss/ (accessed 14 November 2017)

PAGE 60: ...such as exposure to radiation, or in some cases to alcohol...
Juan I Garaycoechea, *et al.*, "Alcohol and Endogenous Aldehydes Damage Chromosomes and Mutate Stem Cells" *Nature*, 03 January 2018, 553, pp. 171–177; and Zhou F C, *et al.*, "Alcohol Alters DNA Methylation Patterns and Inhibits Neural Stem Cell Differentiation" *Alcoholism, Clinical and Experimental Research*, Apr 2011, 35:4, pp. 735-46.

PAGE 62: as L P Hartley told us...
L P Hartley, *The Go-Between* (London: Penguin, 1997), p. 5.

PART II

PAGE 66: Recently, Sally McBrearty and Allison Brooks presented evidence...
Sally McBrearty and Allison Brooks, "The Revolution That Wasn't: A New Interpretation of the Origin of Modern Human Behavior", *Journal of Human Evolution*, December 2000, 39:5, pp.453-563.

PAGE 70: Ohalo is the Pompeii of the Agricultural Revolution...and were capable fishers
See, D Nadel, "Ohalo II: A 23,000-Year-Old Fisher-Hunter-Gatherer's Camp on the Shore of Fluctuating Lake Kinneret (Sea of Galilee)" in Y Enzel & O Bar-Yosef eds, *Quaternary of the Levant: Environments, Climate Change, and Humans* (Cambridge: Cambridge University Press, 2017), pp. 291-294; Steven Mithen, *After the Ice: a Global Human History, 20,000–5,000 BC* (Cambridge, Massachusetts: Harvard University Press, 2006) p. 20; Ehud Weiss, *et al.*, "Plant-food preparation area on an Upper Paleolithic Brush Hut Floor at Ohalo II, Israel" *Journal of Archaeological Science*, 2008, 35, pp. 2400–2414; and, D Nadel, D, *et al.*, From the Cover: Stone Age Hut in Israel Yields World's Oldest Evidence of Bedding, *PNAS*, April 27, 2004, 101: 17, pp. 6821-6826.

CHAPTER 3

PAGE 72: Archaeologists Ainit Snir and Dani Nadel...
Ainit Snir, *et al.*, "The Origin of Cultivation and Proto-

Weeds, Long Before Neolithic Farming" *PLOS One*, July 22, 2015, doi.org/10.1371/journal.pone.0131422 (accessed 7th March 2018).

PAGE 72: Among the techniques used were...a mainstay of many a barbecue today.
See, E C Ellwood, *et al.*, "Stone-Boiling Maize With Limestone: Experimental Results and Implications for Nutrition Among SE Utah Preceramic Groups" *Journal of Archaeological Science*, 2013, 40:1, pp. 35-44; K Nelson, "Environment, Cooking Strategies and Containers", *Journal of Anthropological Archaeology*, 2010, 29:2, pp. 238-247; University of Leiden, "How People Prepared Food in Prehistoric Times" *ScienceDaily*, 16 March 2016, sciencedaily.com/releases/2016/03/160316105806.htm (accessed 10th March 2018); Cathy K Kaufman, *Cooking in Ancient Civilizations* (Westport, Connecticut: Greenwood, 2006); and, A V Thoms "Rocks of Ages: Propagation of Hot-Rock Cookery in Western North America", *Journal of Archaeological Science*, 2009, 36:3, pp. 573-591.

PAGE 73: Some research has even postulated it as the cause of the speculative bump in human IQ...
See, Richard Wrangham, *Catching Fire: How Cooking Made Us Human* (New York: Basic Books, 2010).

PAGE 76: The rates at which dental cavities appear in the fossil record...
Oral Health Foundation, Facts and Figures, http://www.nationalsmilemonth.org/facts-figures/ (accessed 11 December 2017). See also, *Adult Dental Health Survey 1978 and 2009* (England, Wales and Northern Ireland); *The Scottish Health Survey: Volume 1: Main Report*; NHS Dental Epidemiology Programme for England; Oral Health Survey Of Five Year Old Children, 2011/2012; NHS Dental Epidemiology Programme for England; Oral Health Survey of Five Year Old Children, 2007/2008; *Oral Health – Special Eurobarometer* February 2010, 330; British Dental Health Foundation Survey, 2010; and British Dental Health Foundation Survey, 2013.

PAGE 77: A 2014 review article, "Age Changes of Jaws and Soft Tissue Profile"...
Padmaja Sharma, *et al.*, "Age Changes of Jaws and Soft Tissue Profile", *The Scientific World Journal*, 2014

PAGE 78: Harvard-based evolutionary biologist Daniel Lieberman...
Daniel Lieberman, *et al.*, "Effects of Food Processing on Masticatory Strain and Craniofacial Growth in a Retrognathic Face", *Journal of Human Evolution*, June 2004, 46:6, pp. 655-677.

PAGE 78: "Noreen von Cramon-Taubadel at the University of Buffalo New York...
Noreen von Cramon-Taubadel, "Global Human Mandibular Variation Reflects Differences in Agricultural and Hunter-Gatherer Subsistence Strategies", *PNAS* November 21, 2011.

PAGE 78: Another study of the fossil record analysed 292 specimens...
Noreen von Cramon-Taubadel, "Incongruity between Affinity Patterns Based on Mandibular and Lower Dental Dimensions following the Transition to Agriculture in the Near East, Anatolia and Europe", *PLOS One*, 2015.

PAGES 80: There are fossils in China that date...
Prof. Alan Mann, quoted in Douglas Main, "Ancient Mutation Explains Missing Wisdom Teeth", March 13, 2013, www.livescience.com/27529-missing-wisdom-teeth.html (accessed 18 January 2018)

PAGES 81: A study called "The plant component of an Acheulian diet..."
Yoel Melamed, *et al.*, "The plant Component of an Acheulian Diet at Gesher Benot Ya'aqov, Israel", *PNAS*, 2016.

PAGE 82: A genetic study of the bacteria *Streptococcus mutans*...
Cornejo O E, *et al.*, "Evolutionary and Population Genomics of the Cavity Causing Bacteria *Streptococcus mutans*", *Molecular Biology and Evolution Society*, 2012.

PAGE 82: In June 2017, *The Bulletin of the International Association for Paleodontology* published a study of some old remains...
David Frayer, *et al.*, "Prehistoric dentistry? P4 Rotation, Partial M3 Impaction, Toothpick Grooves and Other Signs of Manipulation in Krapina Dental Person 20", *Bulletin of the International Association for Paleodontology*, 2017, 11:1, pp. 1-10.

PAGE 83: A team of researchers led by Stefano Benazzi at the University of Bologna...
Stefano Benazzi, *et al.*, "Earliest Evidence of Dental Caries Manipulation in the Late Upper Palaeolithic", *Scientific Reports* 2015, 5:12150.

PAGE 83: ...a discovery in 2012 by an Italian team found a stone age tooth in Slovenia...
F Bernardini, *et al.*, "Beeswax as Dental Filling on a Neolithic Human Tooth" *PLOS One*, September 19, 2012.

PAGE 83: Ancient Egyptians must have been rubbing their cheeks...
"History of Dentistry", American Dental Education Association, www.adea.org/GoDental/Health_Professions_Advisors/History_of_Dentistry.aspx (accessed 6th August 2017)

PAGE 85: This phenomenon is called the Stephan Curve...
W H Bowen, "The Stephan Curve Revisited", *Odontology*, Jan 2013, 101:1, pp. 2-8.

PAGE 86: A 2013 study, "Fasting in mood disorders: neurobiology and effectiveness"...
G Fond, *et al.*, "Fasting in Mood Disorders: Neurobiology and Effectiveness. A Review of the Literature", *Psychiatry Research*, Oct 30 2013, 209:3, pp. 253-258.

PAGE 86: Longo published a review with Mark Mattson...
V D Longo and M Mattson, "Fasting: Molecular Mechanisms and Clinical Applications", *Cell Metabolism*, Feb 2014, 19:2, pp. 181-192.

PAGE 86: Mattson has also gone on to research meal times...
M Mattson, *et al.*, "Meal Frequency and Timing in Health and Disease", *Proceedings of the National Academy of Sciences*, Nov 2014, 111:47, pp. 16647-53.

PAGE 87: It is early days, but recent research suggests that fasting is also beneficial...
Seyed Mohammad Amin Kormi, *et al.*, "The Effect of Islamic Fasting in Ramadan on Osteoporosis", *Journal of Fasting and Health*, Spring 2017, 5:2, pp. 74-77.

CHAPTER 4

PAGE 92: Today, nearly half the world's population is at risk of malaria...
"Fact Sheet: World Malaria Report 2016" – World Health Organization www.who.int/malaria/media/world-malaria-report-2016/en/ (accessed 8 Jun 2017)

PAGE 92: "A recent study by the University of Central Florida ...
Robert C Sharp, *et al.*, "Polymorphisms in Protein Tyrosine Phosphatase Non-receptor Type 2 and 22 (PTPN2/22) Are Linked to Hyper-Proliferative T-Cells and Susceptibility to Mycobacteria in Rheumatoid Arthritis", *Frontiers in Cellular and Infection Microbiology*, Jan 2018; see also, Robert C Sharp, *et al.*, "Genetic Variations of PTPN2 and PTPN22: Role in the Pathogenesis ofType 1 Diabetes and Crohn's Disease", *Frontiers in Cellular and Infection Microbiology*, December 2015.

PAGE 93: A 2008 study, "Earliest date for milk use...
Richard P Evershed, *et al.*, "Earliest Date for Milk Use in the Near East and Southeastern Europe Linked to Cattle Herding", *Nature*, September 2008, 455, pp. 528–531.

PAGE 93: A study published in the American Journal of Human Genetics in 2006...
E Patin, *et al.*, "Deciphering the Ancient and Complex Evolutionary History of Human Arylamine N-Acetyltransferase Genes", *American Journal of Human Genetics*, Mar 2006, 78:3, pp. 423-36.

PAGE 94: Analysis of the skeletal engravings (*serpens endocrania symmetrica*)...
Israel Hershkovitz, *et al.*, "Detection and Molecular Characterization of 9000-Year-Old Mycobacterium Tuberculosis from a Neolithic Settlement in the Eastern Mediterranean", *Plos One*, 2008.

PAGE 98: According to Michael Hermanussen in his study "Stature in Early Europeans"...
Michael Hermanussen, "Stature in Early Europeans" *Hormones*, 2003, 2:3, pp. 175-178.

PAGE 98: Thanks to the tireless experiments of those early 19th century geneticists...
For a good overview of height, genetics and their entwined history, see chapter 4 "The End of Race" in Adam Rutherford, *A Brief History of Everyone Who Ever Lived: The Stories in Our Genes* (London: Weidenfeld & Nicolson, 2016)

PAGE 99: A study conducted in 2010 used the genetic data of 183,727 individuals...
Hana Lango Allen, *et al.*, "Hundreds of Variants Clustered in Genomic Loci and Biological Pathways Affect Human Height" *Nature*, October 2010, 467, pp. 832–838.

PAGE 100: ...the Sardinian story is one that subscribes to Aristotle's dramatic rules...
See, Aristotle, *The Poetics of Aristotle*, trans. Samuel Henry Butcher (London: Macmillan, 1907), sections VII and VIII.

PAGES 101: Dr Gillebert D'Hercourt was probably the first serious anthropologist...
Gillebert D'Hercourt, *Rapport sur l'anthropologie et l'ethnologie des populations sardes Archives des missions scientifiques et litteraires, tome I* (Paris, 1850).

PAGE 101: A recent study described analytic research into the history of Sardinian stature...
G Pes, *et al.*, "Why Were Sardinians the Shortest Europeans? A Journey Through Genes, Infections, Nutrition, and Sex", *American Journal of Physical Anthropology* May 2017, 163:1, pp. 3-13.

PAGE 103: Barry Bogin, Professor of Biological Anthropology...
See B Bogin, "Secular Changes in Childhood, Adolescent and Adult Stature", *Nestlé Nutrition Institute Workshop Series*, 2013, 71, pp. 115-26; and Agata Blaszczak-Boxe, "Taller, Fatter, Older: How Humans Have Changed in 100 Years" *Live Science*, July 21, 2014, www.livescience.com/46894-how-humans-changed-in-100-years.html (accessed 2 Apr 2017).

PAGE 103: A 2012 study collated data from 10,317 women and men...
Gert Stulp, "Intralocus Sexual Conflict Over Human Height", *Biology Letters*, Dec 2012, 8:6, pp. 976–978.

CHAPTER 5

PAGE 110: Yet, in a collection of *Babylonian Wisdom Literature*...
"My vagina is fine; (yet) according to my people (its use) for me is ended", in Wilfred G Lambert, *Babylonian Wisdom Literature* (Oxford: Oxford University Press, 1960), p. 248.

PAGE 112: It is a common sentiment that Jonathan Swift's *Gulliver's Travels*...
Jonathan Swift, *Gulliver's Travels* (London: Penguin, 2003).

PAGE 113: A famous study of Danish twins...
A M Herskind, *et al.*, "The heritability of human longevity: a population-based study of 2872 Danish twin pairs born 1870-1900", *Human Genetics*, Mar 1996, 97:3, pp. 319-23.

PAGE 114: As Dan Buettner says in his book...
Dan Buettner, *The Blue Zones: Nine Lessons for Living Longer from the People Who've Lived the Longest* (Washington D C:National Geographic Society, 2008).

PAGE 116: A trial by a group of gerontologists based at Boston University...
Stacy L Andersen, *et al.*, "Health Span Approximates Life Span Among Many Supercentenarians: Compression of Morbidity at the Approximate Limit of Life Span", *Journal of Gerontology Series: Biological Sciences and Medical Sciences*, Apr 2012, 67A:4, pp. 395–405.

PAGE 116: Despite our investing hundreds of billions into healthcare...
Office for National Statistics, "An overview of lifestyles and wider characteristics linked to Healthy Life Expectancy in England: June 2017", www.ons.gov.uk/peoplepopulationandcommunity/healthandsocialcare/healthinequalities/articles/healthrelatedlifestylesandwidercharacteristicsofpeoplelivinginareaswiththehighestorlowesthealthylife/june2017 (accessed 28 June 2017).

PAGE 119: Cicero, the Roman politician and lawyer, celebrated the fact that...
See, Cicero, "Cato the Elder on Old Age: On Old Age" in *Cicero Selected Works* trans. Michael Grant (London: Penguin, 1960), pp. 227-8

PAGE 119: Pliny the Younger, a writer and also a Roman lawyer, explained…
Pliny, "Letter to Cornelius Tacitus VI" in *The Letters of the Younger Pliny* trans. Betty Radice (London: Penguin, 2003), p. 39.

PAGE 119: But in 2017 a study was published in the UK which reported…
Research published by Public Health England "Physical inactivity levels in adults aged 40 to 60 in England 2015 to 2016", Published 24 August 2017, www.gov.uk/government/publications/physical-inactivity-levels-in-adults-aged-40-to-60-in-england/physical-inactivity-levels-in-adults-aged-40-to-60-in-england-2015-to-2016 (accessed 12 Jun 2018)

PAGE 120: But Active 10 lowers its expectations…
See https://www.nhs.uk/oneyou/active10/home?utm_source=PR&utm_medium=PR&utm_campaign=Active10 (accessed 12 Jun 2018)

PAGE 124: But a study in 2018 showed that cyclists who remained very active…
N A Duggal, *et al.*, "Major Features of Immunesenescence, Including Reduced Thymic Output, Are Ameliorated by High Levels of Physical Activity in Adulthood", *Aging Cell*, Apr 2018, 17:2.

PAGE 124: Roger Corder, Professor of Experimental Therapeutics at the William Harvey Research Institute at Queen Mary University…
Roger Corder, *The Wine Diet* (London: Sphere, 2009).

PART III

PAGE 130: Working class autobiographers of the 19th century such as William Dodd and Robert Blincoe…
William Dodd, *A Narrative of the Experience and Sufferings of William Dodd a Factory Cripple* (London: 1841); William Dodd, *The Labouring Classes of England* (Boston: 1847); and, John Brown and Robert Blincoe, *A Memoir of Robert Blincoe, an Orphan Boy; Sent from the Workhouse of St Pancras, London, at Seven Years of Age, to Endure the Horrors of a Cotton-Mill Through his Infancy and Youth* (Manchester, 1832).

PAGE 131: In 1837 Charles Wing, surgeon to the Metropolitan Hospital for Children…
Charles Wing, *Evils of the Factory System: Demonstrated by Parliamentary Evidence* (London, 1837).

PAGE 131: A mass observation by surgeons in 1836–7 reported…
"Factory Children" in *Accounts and Papers: Trade and Navigation, Factories, Post Office*, etc., for the *Session 31 January – 17 July* vol 12 of 15, (London, 1837), pp. 1-8.

PAGE 131: James Harrison, a surgeon in Preston, found a mean height of 152cm (5ft)…
See, "Factory Children" in *Accounts and Papers: Trade and Navigation, Factories, Post Office, etc.*, for the *Session 31 January – 17 July* vol 12 of 15, (London, 1837), pp. 1-8; see also, Roderick Floud and Bernard Harris, "Health, Height, and Welfare: Britain, 1700-1980" in *Health and Welfare During Industrialization,* eds Richard H Steckel and Roderick Floud, (Chicago: University of Chicago Press, 1997), pp. 91-126; and, Peter Kirby, *Child Workers and Industrial Health in Britain, 1780-1850* (Woodbridge: Boydell, 2013), p. 112.

PAGE 133: The artist Nickolay Lamm recently studied photos and illustrations of the male body…
See, "Male Body Ideals Through Time", lammily.com/magazine/male-body-ideals-through-time/ (accessed 10 July 2017)

PAGE 133: In 1833, Peter Gaskell produced a study which considered the lot of the pre-industrial worker…
Peter Gaskell, *The Manufacturing Population of England, Its Moral, Social, and Physical Conditions, and the Changes Which Have Arisen from the Uses of Steam Machinery* (London: 1833).

PAGE 136: In Dickens' *Oliver Twist*…He is "one of those long-limbed, knock-kneed, shambling, bony people".
Charles Dickens, *Oliver Twist* (Oxford: Oxford University Press, 1998), p. 397.

PAGE 137: In his book *The Conditions of the Working Class in England*…
Friedrich Engels, *The Condition of the Working Class in England* (London: Penguin, 2009), p. 173.

PAGE 139: This is Anthropocene body pain…
Chandra Prakash Pal, *et al.*, "Epidemiology of knee osteoarthritis in India and related factors", *Indian Journal of Orthopaedics*, Sept 2016, 50:5, pp. 518–522.

PAGE 139: In the UK, according to figures published in 2013…
Osteoarthritis in General Practice – Data and Perspectives (Arthritis Research UK), July 2013.

PAGE 139: In this 2017 research a team at Harvard reported their first findings from a much larger study…
Ian J Wallace, *et al.*, "Knee Osteoarthritis Has Doubled in Prevalence Since the Mid-20th century", *PNAS*, 2017, 114:35, pp. 9332-9336.

PAGE 143: To accentuate the cultural permeation of upholstery in the period…
William Cowper, *The Task: a Poem, in Six Books* (London: 1785).

PAGE 147: The book's famous opening sentence is: "Now, what I want is Facts"…
Charles Dickens, *Hard Times, for These Times* (London: Penguin, 1995), p. 9.

PAGE 149: Environmental psychologists have run trials to assess…
Frances Kuo, *et al.*, "A Potential Natural Treatment for Attention-Deficit/ Hyperactivity Disorder: Evidence from a National Study", *American Journal of Public Health*, Sept 2004, 94:9, pp. 1580-1586

PAGE 150: In 2010, a research paper presented evidence showing that those children who were youngest in their class…
T E Elder, "The Importance of Relative Standards in ADHD Diagnoses: Evidence Based on Exact Birth Dates", *Journal of Health Economics*, Sep 2010, 29:5, pp. 641-56.

CHAPTER 6

PAGE 151: In 1863 Mary Ann Walkley, indirectly employed by the royal court, died of overwork…
Alison Matthews David, *Fashion Victims: The Dangers of Dress Past and Present* (London: Bloomsbury, 2015), p. 8.

PAGE 152: Today, according to *The Lancet*'s "Global Burden of Disease Study of 2015"…GBD 2015 Mortality and Causes of Death Collaborators, "Global, Regional, and National Life Expectancy, All-

Cause Mortality, and Cause-Specific Mortality for 249 Causes of Death, 1980–2015: A Systematic Analysis for the Global Burden of Disease Study 2015", 8 October 2016, 388:10053, pp. 1459–1544.

PAGE 154: A remarkable 2017 study from Deakin University in Australia...
Daniel L Belavý, *et al.*, "Running Exercise Strengthens the Intervertebral Disc", *Scientific Reports*, 2017, 7:45975.

PAGE 159: The idea that there was a "Condition of England Question" was Carlyle's...
Thomas Carlyle, *Chartism* (London, 1839); *Signs of the Times* (London, 1829); *Latter-Day Pamphlets* (London, 1850); *Past and Present* (London, 1843), see, Chapter 4 "Captains of Industry".

PAGE 160: In *The Stones of Venice*...
John Ruskin, *The Stones of Venice*, in Ruskin, *The Works of John Ruskin*, eds Edward Tyas Cook and Alexander Wedderburn, 39 vols. (London: George Allen, 1903–12); see also, John Ruskin, *Seven Lamps of Architecture* (London: George Allen, 1903), p. 218.

PAGE 160: Later, William Morris, an artist and designer who turned increasingly to politics...
William Morris, "Useless Work Versus Useless Toil" in, *News from Nowhere and Other Writings* (London: Penguin, 1993).

PAGE 160: William Dodd in *The Labouring Classes of England* showed...
William Dodd, *The Labouring Classes of England* (Boston: 1847).

PAGE 161: Hippocrates was the earliest writer to work explicitly on ergonomics...
"Ergonomics in Ancient Greece", The Ergonomics Unit – National Technical University of Athens, ergou.simor.ntua.gr/research/ancientGreece/AncientGreece.htm (accessed 19 September 2017)

PAGE 163: One of the earliest exercises in popularizing sometimes erroneous healthcare...
Edward W Duffin, *On Deformities of the Spine* (London, 1848).

PAGE 164: In the late 1850s, the first "study" of ergonomics was published by a Polish physician...
Wojciech Jastrzebowski, *An Outline of Ergonomics or Science of Work based upon truths drawn from the science of nature* (London, 1857).

PAGE 165: Several studies have shown that following an ergonomic intervention...
See, Peter Buckle, *et. al.*, *Work-Related Neck and Upper Limb Musculoskeletal Disorders*, (European Agency for Safety and Health at Work, 1999), especially section 3.4; "Standing Up for Workplace Wellness, a White Paper", Ergotron, 2011, p. 8; and, B Husemann, *et al.*, "Comparisons Of Musculoskeletal Complaints and Data Entry Between a Sitting and a Sit-Stand Workstation Paradigm" *Human Factors: The Journal of Human Factors and Ergonomics Society*, June 2009, 51: 3,. pp. 310-320.

PAGE 169: That philosopher-monk and early man of science Roger Bacon...
Roger Bacon, *The "OPUS MAJUS" of Roger Bacon*, edited by John Henry Bridges (Oxford, 1897).

PAGE 169: These tended to be well-to-do bookish types...
Jane Austen, *Pride & Prejudice* (London: Penguin, 1996); Charlotte Brontë, *Shirley* (London: Penguin, 2006), p. 135.

PAGE 170: The pathology that gathered pace in the 19th century is now dominating parts of Asia...
Eli Dolgin, "The Myopia Boom – Short-Sightedness Is Reaching Epidemic Proportions. Some Scientists Think They Have Found a Reason Why", *Nature*, 18 March 2015, www.nature.com/news/the-myopia-boom-1.17120 (accessed 27 Nov 2017); "Short-sightedness (myopia)", *NHSChoices*, www.nhs.uk/conditions/short-sightedness/ (accessed 27 Nov 2017); Rohit Saxena, *et al.*, "Is Myopia a Public Health Problem in India?" *Indian Journal of Community Medicine*, 2013 38:2, pp. 83–85; "13% Schoolchildren Myopic in India: AIIMS", *The Indian Express*, March 14, 2016 indianexpress.com/article/lifestyle/health/13-schoolchildren-myopic-in-india-aiims/ (accessed 27 Nov 2017); and, Hassan Hashemi, *et al.*, "The Prevalence of Refractive Errors in 5–15 Year-Old Population of Two Underserved Rural Areas of Iran", *Journal of Current Ophthalmology*, September 2017, 29:3, pp. 143–232.

PAGE 170: A 1975 study published in the *Canadian Medical Association Journal*...
RW Morgan, *et al.*, "Inuit myopia: an environmentally induced 'epidemic'?" *Canadian Medical Association Journal*, Mar 1975, 112:5, pp. 575–577.

PAGE 170: A study by the University of Cardiff in 2015...
J A Guggenheim, *et al.*, "Role of Educational Exposure in the Association Between Myopia and Birth Order" in *JAMA Ophthalmology*, 2015, 133:12, pp. 1408-14.

PAGE 170: Other studies confound this data when children from similar genetic backgrounds...
K A Rose, *et al.*, "Myopia, lifestyle, and schooling in students of Chinese ethnicity in Singapore and Sydney", *Archives of Ophthalmology*, 2008, 126:4, pp. 527-30.

PAGE 171: Similar studies by the same team dovetailed with these findings...
K A Rose, *et al.*, "Outdoor Activity Reduces the Prevalence of Myopia in Children" in *Ophthalmology*, 2008, 115:8, pp. 1279-85.

PAGE 171: Research done as long ago as the 1930s...
See, S Holm, "The ocular refraction state of the Palae-Negroids in Gabon, French Equatorial Africa" *Acta Ophthalmologica Supplementum* 1937, 13: pp. 1–299; Loren Cordain, *et al.*, "An evolutionary analysis of the aetiology and pathogenesis of juvenile-onset myopia", *Acta Ophthalmologica Supplementum*, April 2002, 80: 2, pp. 125–135; and, D S London *et al.*, "A Phytochemical-Rich Diet May Explain the Absence of Age-Related Decline in Visual Acuity of Amazonian Hunter-Gatherers in Ecuador", *Nutrition Research*, 2015, 35:2, pp. 107-17.

PAGE 172: In 2016 in the UK, a report was published that concluded that 74 percent of British children suffered lack of access to green spaces...
Damian Carrington, "Three-quarters of UK Children Spend Less Time Outdoors Than Prison Inmates – Survey", *The Guardian*, www.theguardian.com/environment/2016/mar/25/three-quarters-of-uk-children-spend-less-time-outdoors-than-prison-inmates-survey (accessed 13 Mar 2017)

PAGE 172: According to another government report published in 2016…
"Monitor of Engagement with the Natural Environment Pilot Study: Visits to the Natural Environment by Children", *Natural England Commissioned Report NECR208*, 10 February 2016; see also, Patrick Barkham and Jessica Aldred, "Concerns Raised Over Number of Children Not Engaging With Nature", *The Guardian*, www.theguardian.com/environment/2016/feb/10/concerns-raised-over-amount-of-children-not-engaging-with-nature (accessed 13 Mar 2017)

PAGE 172: Using data from 145 studies involving 2.1 million participants…
B A Holden, *et al.*, "Global Prevalence of Myopia and High Myopia and Temporal Trends from 2000 through 2050", *Ophthalmology*, May 2016, 123:5, pp. 1036-42.

CHAPTER 7

PAGE 174: The prophet of 19th-century population growth was Thomas Malthus…
Thomas Malthus, *Essay on the Principle of Population* (London, 1798).

PAGE 177: London required the felling of approximately 120,000 trees…
In 1919, forest management was in such disarray that it became a national concern and saw the inauguration of the Forestry Commission. See, Oliver Rackham, *Woodlands* (London: Collins, 2015), especially Chapter 3; Oliver Rackham, Ancient Woodland: Its History, Vegetation and Uses in England (Castlepoint, 2003); "The State of the UK's Forest, Woods and Trees: Perspectives from the Sector", *The Woodland Trust* 2011; "What Shaped Britain's Forests?", *Forestry Commission*, https://www.forestry.gov.uk/forestry/INFD-8Y5BSY (accessed 4 Aug 2017); Bibi van der Zee, "England's forests: a brief history of trees", *The Guardian*, www.theguardian.com/travel/2013/jul/27/history-of-englands-forests (accessed 4 Aug 2017)

PAGE 177: In 1661, the famous diarist and fellow of the Royal Society, John Evelyn…
John Evelyn, *Fumifugium: or The Inconveniencie of the Aer and Smoak of London Dissipated* (London, 1661).

PAGE 177: John Graunt, a fellow of the Royal Society…
John Graunt, *Natural and Political Observations Mentioned in a following Index and made upon the Bills of Mortality… with reference to the Government, Religion, Trade, Growth, Air, Diseases and the several Changes of the said City*, 1676, in William Petty, *The Economic Writings of Sir William Petty*, ed. Charles Henry Hull (Cambridge, 1899), 2, pp. 393-94.

PAGE 177: Between the time of the Domesday Book and the mid-14th century…
Josiah Cox Russell, *British Medieval Population*, (Alberqueque, University of New Mexico Press, 1948), pp. 263, 269; see also, John Graunt, "Natural and Political Observations Mentioned in a Following Index, and Made Upon the Bills of Mortality", in *Mathematical Demography*, eds David P Smith Nathan Keyfitz (Springer-Verlag: Berlin, 1977), pp. 11–20.

PAGE 178: In 1850, for example, a German mine might have produced 8,000–9,000 tonnes …
Emma Griffin, "Why was Britain First? The Industrial Revolution in Global Context", *Short History of the British Industrial Revolution* (London: Palgrave, 2010), p. 163-5.

PAGE 178: France boasted similar figures…
Emma Griffin, "Why was Britain First? The Industrial Revolution in Global Context", *Short History of the British Industrial Revolution* (London: Palgrave, 2010), p. 168.

PAGE 179: By the dawn of the 20th century, the respiratory disease bronchitis was Britain's biggest killer…
Stephen Mosley, quoted in, Jim Morrison, "Air Pollution Goes Back Way Further Than You Think" www.smithsonianmag.com/science-nature/air-pollution-goes-back-way-further-you-think-180957716/#6Y7FJM1kDE2xzm6l.99 (accessed 25 Aug 2017).

PAGE 179: Bostock claimed he had scoured the country to find sufferers…
J Bostock, "Of the Catarrhus Æstivus, or Summer Catarrh", *Medico-Chirurgical Transactions*, 1828, 14:2, pp. 437–446.

PAGE 179: One of the first times it was mentioned by name was in Benjamin Disraeli's novel…
Benjamin Disraeli novel, *Sybil, or the Two Nations* (London, 1845).

PAGE 180: Charles Blackley first announced a causal link between pollen and the disease in 1873.
Charles Blackley, *Hay Fever: Its Causes, Treatment, and Effective Prevention Experimental Researches* (London, 1880).

PAGE 180: Morell Mackenzie, a British physician in the 1880s…
Morell Mackenzie, *Hay Fever and Paroxysmal Sneezing: Their Etiology and Treatment – an Appendix on Rose Cold* (London, 1887).

PAGE 180: E M Forster's *Howards End*…
E M Forster, *Howards End* (London: Edward Arnold, 1910).

PART IV

PAGE 191: In 1861, the census suggests that about 91,000 people…
See, *1861 Census of England and Wales, General Report; with appendix of tables* (1863 LIII (3221) 1); and, *Census of England and Wales, 1891, Ages, condition as to marriage, occupations, birth-places and infirmities*, Vol. III BPP 1893–4 CVI.

PAGE 197: A study published in the *British Journal of Sports Medicine* in 2014 found…
A E Staiano, *et al.*, "Sitting Time and Cardiometabolic Risk in US Adults: Associations by Sex, Race, Socioeconomic Status and Activity Level", *British Journal of Sports Medicine* in Feb 2014, 48:3, pp. 213-9.

PAGE 198: "Another study, published in the *American Journal of Epidemiology* in 2010…
A V Patel, *et al.*, "Leisure Time Spent Sitting in Relation to Total Mortality in a Prospective Cohort of US Adults", *American Journal of Epidemiology*, Aug 2010, 172:4, pp. 419-29.

PAGE 198: A study published in the journal *Diabetes* in 2007…
M T Hamilton, *et al.*, "Role of Low Energy Expenditure and Sitting in Obesity, Metabolic Syndrome, Type 2

Diabetes, and Cardiovascular Disease", Nov 2007, 56:11, pp. 2655–67.

PAGE 198: This was echoed in a 2010 study at the University of Queensland's School of Occupational Health...
N Owen, *et al.*, "Too Much Sitting: A Novel and Important Predictor of Chronic Disease Risk?", *British Journal of Sports Medicine*, 2009, 43, pp. 81–83.

PAGE 199: A 2009 study at the University of South Carolina...
TY Warren, *et al.*, "Sedentary Behaviors Increase Risk of Cardiovascular Disease Mortality in Men", *Medicine and Science in Sports and Exercise*, May 2010, 42: 5, pp. 879–85.

PAGE 199: The 2010 American Cancer Society study...
A V Patel, *et al.*, "Leisure time spent sitting in relation to total mortality in a prospective cohort of US adults", *American Journal of Epidemiology*, August 2010, 172:4, pp. 419–29.

PAGE 200: A 2012 study published in the *American Journal of Epidemiology*...
M Du, *et al.*, "Physical Activity, Sedentary Behavior, and Leukocyte Telomere Length in Women", *American Journal of Epidemiology*, March 2012, 175:5, pp. 414–22.

PAGE 201: A trial conducted with 480 schoolchildren and 25 teachers...
M E Benden, *et al.*, "The Evaluation of the Impact of a Stand-Biased Desk on Energy Expenditure and Physical Activity for Elementary School Students", *International Journal of Environmental Research and Public Health*, Sep 2014, 11: 9, pp. 9361–9375.

PAGE 202: A massive review which collated the data from 1,827 previous trials concluded...
S A Clemes, *et al.*, "What Constitutes Effective Manual Handling Training?", *Occupational Medicine*, Mar 2010, 60:2, pp. 101–107.

PAGE 204: Early in 2018, an article in the *European Journal of Preventive Cardiology*...
F Saeidifard, *et al.*, "Differences of Energy Expenditure While Sitting Versus Standing: A Systematic Review and Meta-Analysis", *European Journal of Preventive Cardiology*, Jan 2018, 25:5, pp. 522–538.

PAGE 204: A recent study from Curtin University in Australia has found that...
R Baker, *et al.*, "A Detailed Description of the Short-Term Musculoskeletal and Cognitive Effects of Prolonged Standing for Office Computer Work", *Ergonomics*, 2018, 61:7, pp. 877...890.

PAGE 205: A study by Ivan Lin focused on a population that had previously seemed to be insulated from the effects of chronic low back pain...
I B Lin, *et al.*, "Disabling Chronic Low Back Pain as an Iatrogenic Disorder: A Qualitative Study in Aboriginal Australians", *BMJ Open*, 2013, 3:4.

PAGE 209: Research published in 2016 by biomechanics expert Hannah Rice...
H M Rice, *et al.*, "Footwear Matters: Influence of Footwear and Foot Strike on Load Rates During Running", *Journal of Medicine & Science in Sport & Exercise*, Dec 2016, 48:12, pp. 2462–2468.

PAGE 209: But in the last century or so they have started getting bigger...
J Laurance, "Why are Our Feet Getting Bigger", *Independent*, Tuesday 3 June 2014, www.independent.co.uk/life-style/health-and-families/features/why-our-feet-are-getting-bigger-9481529.html (accessed 13 Oct 2017)

PAGE 209: A 2017 study of 500 Indian women and men found...
See A Aenumulapalli, *et al.*, "Prevalence of Flexible Flat Foot in Adults: A Cross-sectional Study, *Journal of Clinical and Diagnostic Research*", Jun 2017, 11:6, AC17–AC20; also U B Rao, *et al.*, "The influence of Footwear on the Prevalence of Flat Foot", Journal of Bone & Joint Surgery, 1992; and N B Holowka, *et al.*, "Foot Strength and Stiffness Are Related to Footwear Use in a Comparison of Minimally – vs. Conventionally-shod Populations", *Scientific Reports*, 2018, 8: 3679.

PAGE 209: Another 2017 study found that "habitual footwear use has significant effects..."
K Hollander, *et al.*,"Growing-up (Habitually) Barefoot Influences the Development of Foot and Arch Morphology in Children and Adolescents", *Nature: Scientific Reports*, 7:1, p. 8079.

CHAPTER 8

PAGE 216: According to the Office for National Statistics...
"Life Expectancy at Birth and at Age 65 by Local Areas in England and Wales", Office for National Statistics, 4 Nov 2015, www.ons.gov.uk/peoplepopulationandcommunity/birthsdeathsandmarriages/lifeexpectancies/datasets/lifeexpectancyatbirthandatage65bylocalareasinenglandandwalesreferencetable1; "Life Expectancy Rises 'Grinding to Halt'", 19 July 2017, www.ucl.ac.uk/iehc/iehc-news/michael-marmot-life-expectancy, (accessed 25 Sep 2017).

PAGE 220: His book, entitled *Degeneration: a Chapter on Darwinism*
Edwin Ray Lankester, *Degeneration: a Chapter on Darwinism* (London, 1880), see pp. 26–32.

PAGE 221: In research recently published by Public Health England the authors found that about one in two women and a third of men in England are damaging their health through lack of physical activity...
J Varney, *et al.*, "Everybody Active, Every Day: Two Years On: an Update on the National and Physical Activity Framework" *Public Health England*, 2017.

PAGE 222: As things stand, obesity and its related diseases are estimated to be likely to cost the global economy $1.2 trillion each year by 2025
"Avoiding the Consequences of Obesity", World Obesity Federation, www.obesityday.worldobesity.org/world-obesity-day-2017 ; see also "Our Data" http://www.obesityday.worldobesity.org/ourdata2017 (accessed 10 December 2017).

PAGE 222: According to the US National Center for Health Statistics, a chronic disease is one lasting three months or more, cannot be prevented by vaccines or cured by medications, nor does it just disappear.
N J Farpour-Lambert, *et al.*, "Childhood Obesity Is a Chronic Disease Demanding Specific Health Care – a

Position Statement from the Childhood Obesity Task Force (COTF) of the European Association for the Study of Obesity (EASO)", *Obesity Facts*, 2015, 8:5, pp. 342–9.

PAGE 223: Research published in *Science* in 2013 suggested just this
V K Ridaura, *et al.*, "Gut Microbiota from Twins Discordant for Obesity Modulate Metabolism in Mice" *Science*, Sep 2013 6:341.

PAGE 224: A Stanford-based study in 2018 found just this. People who cut back on processed foods...
C D Gardner, *et al.*, "Effect of Low-Fat vs Low-Carbohydrate Diet on 12-Month Weight Loss in Overweight Adults and the Association with Genotype Pattern or Insulin Secretion: The DIETFITS Randomized Clinical Trial", *JAMA*, Feb 2018 20:319, pp. 667-679.

PAGE 224: A trial of gut microbiota transfer...
J R Kelly, *et al.*, "Transferring the Blues: Depression-Associated Gut Microbiota Induces Neurobehavioural Changes in the Rat", *Journal Psychiatric Research*, Nov 2016, 82, pp. 109–18.

PAGE 225: To the best of my knowledge, it was a 2012 article in *The Lancet*
Chi Pang Wen, *et al.*, "Stressing Harms of Physical Inactivity to Promote Exercise", *The Lancet*, July 2012, 380: 9838, pp. 192–193.

PAGE 225: Here is a ranking of causes for "disability-adjusted life years"...
GBD 2015 Mortality and Causes of Death Collaborators, "Global, Regional, and National Life Expectancy, All-Cause Mortality, and Cause-Specific Mortality for 249 Causes of Death, 1980–2015: A Systematic Analysis for the Global Burden of Disease Study 2015", *The Lancet*, October 2016, 388:10053, pp. 1459–1544.

PAGE 226: A 2014 study published in *JAMA*...
D Dabelea, *et al.*, "Prevalence of Type 1 and Type 2 Diabetes Among Children and Adolescents from 2001 to 2009", *JAMA*, May 2014 7:311, pp. 1778–86.

PAGE 226: Another study published in *JAMA Paediatrics* in 2014...
S M Virtanen, *et al.*, "Microbial Exposure in Infancy and Subsequent Appearance of Type 1 Diabetes Mellitus-Associated Autoantibodies: A Cohort Study", *JAMA Paediatrics*, Aug 2014, 168:8, pp. 755–63.

PAGE 227: The World Health Organization provides shocking advice on processed meats...
World Health Organization, "Q&A on the carcinogenicity of the consumption of red meat and processed meat", www.who.int/features/qa/cancer-red-meat/en/ (accessed 10 June 2017)

PAGE 228: A new pill will soon be up for FDA....
World Anti-Doping Agency, "WADA issues alert on GW501516", www.wada-ama.org/en/media/news/2013-03/wada-issues-alert-on-gw501516 (accessed 8 Aug 2017)

PAGE 229: A recent trial run by Imperial College London and Harvard's School of Public Health...
NCD Risk Factor Collaboration, "Worldwide Trends in Diabetes Since 1980: A Pooled Analysis of 751 Population-Based Studies With 4·4 Million Participants", *Lancet*, 2016, 387, pp. 1513–30.

PAGE 231: It is for this reason that a Republican representative from Texas, Lamar Smith...
Lamar Smith, "Don't Believe the Hysteria Over Carbon Dioxide", www.dailysignal.com/2017/07/24/dont-believe-hysteria-carbon-dioxide/ (accessed 1 September 2017)

PAGE 231: I recently heard of a study compiled by a mathematician, Irakli Loladze...
Irakli Loladze, "Hidden Shift of the Ionome of Plants Exposed to Elevated CO_2 Depletes Minerals at the Base of Human Nutrition", *eLife*, 2014, 3, e02245.

PAGE 232: In 2002, Loladze published a paper on ecological stoichiometry...
Irakli Loladze, "Rising atmospheric CO_2 and human nutrition: toward globally imbalanced plant stoichiometry?", *Trends in Ecology & Evolution*, Oct 2002, 17:10, pp. 457–461.

PAGE 233: Loladze had an empirical proof that rising CO_2 consistently lowers the quality of plants...
Irakli Loladze, "Hidden Shift of the Ionome of Plants Exposed to Elevated CO_2 Depletes Minerals at the Base of Human Nutrition", *eLife*, 2014, 3, e02245.

PAGE 233: Some of the other uncomfortable science that he is utilizing to assemble his case are studies such as one by Kemnitz...
Klementidis Y C, *et al.*, "Canaries in the Coal Mine: A Cross-Species Analysis of the Plurality of Obesity Epidemics", *Proceedings Biological Sciences and the Royal Society*, Jun 2011, 7: 278, pp. 1626...32.

PAGE 234: A mathematical model developed by Hill...
J O Hill, *et al.*, "Obesity and the Environment: Where Do We Go From Here?", *Science*, Feb 2003 7:299, pp. 853...5.

CHAPTER 9

PAGE 239: An international team of researchers (some from Texas A&M University)...
Gehui Wang, *et al.*, "Persistent Sulfate Formation from London Fog to Chinese haze", *PNAS*, November 2016, 113: 48, pp. 13630...13635.

PAGE 239: Jean Baptiste van Helmont (1579–1644) a physician in Belgium...
L F Haas, "Jean Baptiste van Helmont (1577–1644)", *Journal of Neurology, Neurosurgery & Psychiatry*, 1998, 65:916.

PAGE 240: He found that the workplaces had two major effects...
Bernardino Ramazzini, *Treatise on the Diseases of Workers*, (London, 1705). See also, Bernardino Ramazzini, "De Morbis Artificum Diatriba [Diseases of Workers]", *American Journal of Public Health*, Sep 2001, 91:9, pp. 1380–1382.

PAGE 241: A review article published in *Immunological Reviews* in 2011 ...
Carole Ober and Tsung-Chieh Yao, "The Genetics of Asthma and Allergic Disease: a 21st Century Perspective", *Immunological Reviews*, Jul 2011, 242:1, pp. 10–30.

PAGE 241: When the research was initially published in the *British Medical Journal* in 1989...
D P Strachan, "Hay Fever, Hygiene, and Household Size", *British Medical Journal*, Nov 1989, 299, p. 1259.

PAGE 242: The importance of the natural environment and microbial exposure...
M M Stein, *et al.*, "Innate Immunity and Asthma Risk in Amish and Hutterite Farm Children", *The New England Journal of Medicine*, 2016, 375, pp. 411-421. See also, B A Justyna Gozdz, *et al.*, "Innate Immunity and Asthma Risk", *The New England Journal of Medicine*, 2016, 375, pp. 1898–1899.

PAGE 242: A study published in 2015 explored how the allergy rates of two different sets of people...
T Haahtela, *et al.*, "Hunt for the Origin of Allergy – Comparing the Finnish and Russian Karelia", *Clinical and Experimental Allergy*, May 2015, 5, pp. 891–901.

PAGE 243: A Norwegian study of 1.75 million...
M C Tollånes, *et al.*, "Cesarean Section and Risk of Severe Childhood Asthma: A Population-Based Cohort Study", *Journal of Pediatrics*, July 2008, 153:1, pp. 112–6.

PAGE 245: During the 19th century, Charles Blackley observed...
Charles Blackley, *Hay fever: its Causes, Treatment, and Effective Prevention Experimental Researches* (London, 1880).

PAGE 246: It makes sense then that the numbers are further confirmed by a study...
J Lelieveld, *et al.*, "The Contribution of Outdoor Air Pollution Sources to Premature Mortality on a Global Scale", *Nature*, Sep 2015, 525, pp. 367–371.

PAGE 246: There is little doubt, too, that air quality is the chief driver...
Davies Adeloye, *et al.*, "Global and Regional Estimates of COPD Prevalence: Systematic Review and Meta–analysis", *Journal of Global Health*, Dec 2015, 5:2; and, World Health Organization, "COPD predicted to be third leading cause of death in 2030" www.who.int/respiratory/copd/World_Health_Statistics_2008/en/ (accessed 13 December, 2017)

PAGE 247: The World Health Organization Mortality Database of 2014 reports...
World Health Organization, "WHO Mortality Database", www.who.int/healthinfo/mortality_data/en/ (accessed 15 December, 2017); and, Asthma UK, "Asthma facts and statistics", www.asthma.org.uk/about/media/facts-and-statistics/ (accessed 15 December, 2017).

PAGE 250: In 2012 there were press reports about the surrounding streets being hosed down...
Adam Vaughan, "Boris Johnson's sticky pollution solution shown to be a £1.4m failure. Study finds mayor's trials to 'glue' pollution from vehicles to London's roads have not reduced fine particulate pollution", *The Guardian*, www.theguardian.com/environment/2013/feb/13/boris-johnson-sticky-pollution-failure (accessed 10 December 2017); *The Telegraph*, www.telegraph.co.uk/earth/earthnews/9176543/Boris-accused-of-using-dust-suppressants-to-hide-pollution-levels-before-Olympics.html (accessed 1 August 2014). See also, James Holloway "Pollution suppression in London: flawed or fraud? Is London keeping air pollution down, or simply holding EU regulators at bay?" arstechnica.com/tech-policy/2012/04/the-campaign-for-clean-air/ (accessed 10 December 2017)

PAGE 250: As many as 9,500 people die in London...
Adam Vaughan, "Nearly 9,500 people die each year in London because of air pollution – study. Counting impact of toxic gas NO_2 for the first time suggests more than twice as many people as previously thought die prematurely from pollution in UK capital", *The Guardian*, https://www.theguardian.com/environment/2015/jul/15/nearly-9500-people-die-each-year-in-london-because-of-air-pollution-study ; see also, Richard W Atkinson, *et al.*, "Short-Term Exposure to Traffic-Related Air Pollution and Daily Mortality in London, UK", *Journal of Exposure Science and Environmental Epidemiology*, 2016, 26, pp. 125–132.

PAGE 250: The particles are so small that they are suspended in mid-air...
Air Quality Expert Group, "Fine Particulate Matter (PM2.5) in the United Kingdom - Prepared for: Department for Environment, Food and Rural Affairs; Scottish Executive; Welsh Government; and Department of the Environment in Northern Ireland", 20 Dec 2012.

PAGE 251: Data from DEFRA (Department for Environment, Food & Rural Affairs) show that...
DEFRA, "Interactive monitoring networks map", uk-air.defra.gov.uk/interactive-map

PAGE 251: A study by more than 50 scientists...
John N Newton, *et al.*, "Changes in Health in England, With Analysis by English Regions and Areas of Deprivation, 1990–2013: A Systematic Analysis for the Global Burden of Disease Study 2013", *The Lancet*, December 2015, 386:10010, pp. 2257–2274.

PAGE 252: A recent report by the World Health Organization tells us that up to 92 percent of the world's population is living in a poisonous atmosphere.
World Health Organization, "WHO releases country estimates on air pollution exposure and health impact" www.who.int/news-room/detail/27-09-2016-who-releases-country-estimates-on-air-pollution-exposure-and-health-impact

PAGE 252: A trial published in *Psychological Science* in 2018 examined the ways in which air pollution negatively influenced ethical and criminal behaviour...
J G Lu, *et al.*, "Polluted Morality: Air Pollution Predicts Criminal Activity and Unethical Behavior", *Psychological Science*, Mar 2018, 29:3, pp. 340–355.

PAGE 253: ...Curbing car production would put a £35 billion hole in the Exchequer's pocket.
Office for Budget Responsibility, "Fuel duties", obr.uk/forecasts-in-depth/tax-by-tax-spend-by-spend/fuel-duties/ (accessed 14 Jun 2018); and, "UK car sales at record high in 2016", www.bbc.co.uk/news/business-38516247 (accessed 24 Jan 2018)

PAGE 254: A report published in *JAMA Psychiatry* in 2015 presented some shocking evidence of this...
B S Peterson, *et al.*, "Effects of Prenatal Exposure to Air Pollutants (polycyclic aromatic hydrocarbons) on the Development of Brain White Matter, Cognition, and Behavior in Later Childhood", *JAMA Psychiatry*, Jun 2015, 72:6, pp. 531–40.

PAGE 257: A meta-analysis published in 2011...
Catrine Tudor-Locke, *et al.*, "How Many Steps/Day Are Enough? For Adults" *International Journal of Behavioural Nutrition and Physical Activity*, 2011, 8:79.

PAGE 259: A study published in 2018 suggested that astronauts...

J F Bailey, *et al.*, "From the International Space Station to the Clinic: How Prolonged Unloading May Disrupt Lumbar Spine Stability", *The Spine Journal*, Jan 2018, 18:1, pp. 7–14.

PAGE 259: Beware though; sudden increases in any kind of training are strongly associated with injury.
F Wilson, *et al.*, "A 12-month Prospective Cohort Study of Injury in International Rowers", *British Journal of Sports Medicine*, 2010, 44, pp. 207–214.

PART V

PAGE 265: Hands are about as important to our evolution as sweat and brain size
See, Vybarr Cregan-Reid, "From perspiration to world domination – the extraordinary science of sweat" The Conversation, (theconversation.com/from-perspiration-to-world-domination-the-extraordinary-science-of-sweat-62753).

CHAPTER 10

PAGE 267: In Jill Cook's book...
Jill Cook, *Ice Age Art: the Arrival of the Modern Mind* (London: British Museum Press, 2013).

PAGE 269: ...In the UK in 2018, data tells us that 99 percent of people aged 16-34 are digital natives.
See, International Telecommunications Union, "ICT Facts and Figures", https://www.itu.int/en/ITU-D/Statistics/Pages/stat/default.aspx (accessed 14 Jun 2018); "Countries with the highest number of internet users as of June 2017 (in millions)", www.statista.com/statistics/262966/number-of-internet-users-in-selected-countries/; International Telecommunications Union, "ICT Facts and Figures 2017", www.itu.int/en/ITU-D/Statistics/Documents/facts/ICTFactsFigures2017.pdf; and, "Statistical bulletin: Internet users, UK: 2018", Office for National Statistics, www.ons.gov.uk/businessindustryandtrade/itandinternetindustry/bulletins/internetusers/2018.

PAGE 271: In the past, RSIs were rare...
See, RSI.org, *RSI Awareness – Upper Limb Disorders: an Overview*, rsi.org.uk/pdf/ULDs_Overview.pdf (accessed 26 October 2017); "No progress over RSI injuries", news.bbc.co.uk/1/hi/health/7889091.stm (accessed 26 October 2017); and, Pamela Brown, "Britain counts the cost of RSI and backache", thetimes.co.uk/article/britain-counts-the-cost-of-rsi-and-backache-q5kwv7b6knx (accessed 26 October 2017).

PAGE 9: Across the board, women are affected by the disease more than men...
Peter Buckle and Jason Devereux, "Work-Related Neck and Upper Limb Musculoskeletal Disorders", *Applied Ergonomics*, May 2002, 33:3, pp. 207–17.

PAGE 271: Arthur Munby was a poet...
Arthur Munby, 21st August 1860, in *Derek Hudson, Munby, Man of Two Worlds: the Life and Diaries of Arthur J Munby*, 1812-1910 (Cambridge: Gambit, 1974), p. 71

PAGE 273: If William Dodd is to be believed...
William Dodd, *The Labouring Classes of England* (Boston: 1847), p.116.

PAGE 275: The latest figures published by the Health and Safety Executive..."
Health and Safety Executive, "Work-related Musculoskeletal Disorders (WRMSDs) Statistics in Great Britain 2017", (from, www.hse.gov.uk/statistics/); see also www.rsiaction.org.uk (accessed 14 Jun 2018); and, Lisa Salmon, "Why Repetitive Strain Injury is on the rise", www.irishnews.com/lifestyle/2016/01/27/news/rsi-becoming-increasingly-common-experts-warn-389271/ ((accessed 14 Jun 2018).

PAGE 275: The Chartered Society of Physiotherapy in the UK...
Chartered Society of Physiotherapy, "Sharp rise in rates of repetitive strain injury – physiotherapists call for urgent action by government and employers", www.csp.org.uk/press-releases/2008/02/26/sharp-rise-rates-repetitive-strain-injury-physiotherapists-call-urgent-act (accessed 26 October 2017)

PAGE 281: This piece of research by Pierre Lemelin and Daniel Schmitt...
Pierre Lemelin and Daniel Schmitt, "On Primitiveness, Prehensility, and Opposability of the Primate Hand: The Contributions of Frederic Wood Jones and John Russell Napier" in Tracy L Kivell, *et al.*, *The Evolution of the Primate Hand: Anatomical, Developmental, Functional, and Paleontological Evidence* (Springer: New York, 2016), pp. 5–13. See also, F Wood Jones, Man's Place Among the Mammals (London: Edward Arnold, 1929).

EPILOGUE

PAGE 285: A study that recently appeared in the *British Journal of Sports Medicine*...
S F M Chastin, *et al.*, "How Does Light-Intensity Physical Activity Associate With Adult Cardiometabolic Health and Mortality? Systematic Review With Meta-Analysis of Experimental and Observational Studies", *British Journal of Sports Medicine*, Apr 2018, 25.

PAGE 287 Geneticist Adam Rutherford explains...
Adam Rutherford, *A Brief History of Everyone Who Ever Lived: the Stories in Our Genes* (London: Weidenfeld & Nicolson, 2016).

PAGE 287: There is a wonderful, little-known Christmas book by Charles Dickens...
Charles Dickens, *The Haunted Man, or the Ghost's Bargain* (London, 1848).

PAGE 289: According to research conducted by the British Heart Foundation...
British Heart Foundation, *Physical Inactivity and Sedentary Behaviour Report 2017*

PAGE 290: Currently, more than half the adults in the UK take prescription medication...
Laura Donnelly and Patrick Scott, "Pill nation: half of us take at least one prescription drug daily", www.telegraph.co.uk/news/2017/12/13/pill-nation-half-us-take-least-one-prescription-drug-daily/; and, Mayo Clinic, "Nearly 7 in 10 Americans Take Prescription Drugs, Mayo Clinic, Olmsted Medical Center Find", newsnetwork.mayoclinic.org/discussion/nearly-7-in-10-americans-take-prescription-drugs-mayo-clinic-olmsted-medical-center-find/ (accessed 15 May 2016).

PAGE 291: The World Health Organization estimates that our annual healthcare bill amounts to...

“Global Health Observatory (GHO) data”, www.who.int/gho/health_financing/en/

PAGE 292: A book called *The Fable of the Bees: or, Private Vices, Public Benefits…*
Bernard de Mandeville, *The Fable of the Bees: or, Private Vices, Public Benefits* (London, 1705).

PAGE 293: One publication was…
Bradshaw's Handbook to the Manufacturing Districts (London, 1854).

PAGE 294: A study called “Are vacations good for your health?”
B B Gump and K A Matthews, “Are Vacations Good for Your Health? The 9-Year Mortality Experience After the Multiple Risk Factor Intervention Trial”, *Psychosomatic Medicine* Sep-Oct 2000, 62:5, pp. 608–12.

PAGE 294: The American Institute of Stress has estimated that the condition costs…
American Institute of Stress, “Workplace Stress”, www.stress.org/workplace-stress/ (accessed 26 Apr 2018).

PAGE 294: The United States Department of Labor warns…
United States Department of Labor, “Workplace Violence”, www.osha.gov/SLTC/workplaceviolence/, (accessed 26 Apr 2018).

PAGE 295: In 2010, the *American Journal of Preventative Medicine* published a study of 5,556 people…
A C King, *et al.*, “Identifying Subgroups of US Adults at Risk for Prolonged Television Viewing to Inform Program Development”, *American Journal of Preventive Medicine*, Jan 2010, 38:1, pp. 17–26.

PAGE 297: The Black Dog Institute in Australia…
Samuel B Harvey, *et al.*, “Exercise and the Prevention of Depression: Results of the HUNT Cohort Study”, *American Journal of Psychiatry*, October 2017, 175:1, pp. 28-36.

PAGE 297: A trial run by environmental scientist Mike Rogerson at the University of Essex…
Mike Rogerson, *et al.*, “The Health and Wellbeing Impacts of Volunteering with The Wildlife Trusts”, Report for The Wildlife Trusts, 2017.

PAGE 297: In 2003, Nader Rifai and Paul Ridker…
N Rifai and P M Ridker, “Population Distributions of C-Reactive Protein in Apparently Healthy Men and Women in the United States: Implication for Clinical Interpretation”, *Clinical Chemistry*, Apr 2003, 49:4, pp. 666–9.

PAGE 298: The findings documented “a pattern of variation over time…”
Thomas W McDade, *et al.*, “Analysis Of Variability of High Sensitivity C-Reactive Protein in Lowland Ecuador Reveals No Evidence of Chronic Low-Grade Inflammation”. *American Journal of Human Biology* 5: pp., 675-81. See also, Thomas W McDade, *et al.*, “Population Differences in Associations Between C-Reactive Protein Concentration and Adiposity: Comparison of Young Adults in the Philippines and the United States” *American Journal of Clinical Nutrition*, Apr 2009, 89:4, pp. 1237–1245; and Thomas W McDade, “Early Environments and the Ecology of Inflammation, *PNAS*, Oct 2012, 16:109, pp. 17281–17288.

PAGE 299: Estimates vary, but…
The UN's Department of Economic and Social Affairs in “World population projected to reach 9.7 billion by 2050” estimates the population by 2100 as being 11.2 billion. See, www.un.org/en/development/desa/news/population/2015-report.html (accessed 15 Jun 2018)

INDEX

I

J

K

L

ACKNOWLEDGEMENTS

It is a cliché to thank an editor in the acknowledgements, and to say that they championed the book from its very earliest stages through to completion, but Romilly Morgan redefines what this means; patient, tireless and energetic in equal measure – I'd compare her to a bold and refined heroine in a 19th-century novel, but she'd probably cross that out! Many thanks to you. It has been a pleasure working with everyone at Octopus, particularly Jack Storey, Polly Poulter, Caroline Brown and Matt Grindon, all of whom have been very enthusiastic about the book from the off.

Thanks to my colleagues at the University of Kent for granting me a period of leave in which to get a draft together. For drawing upon their reserves of generosity, thanks go to Eleanor Adkins, Prof Jennie Batchelor (I don't know how she does it!), Dr Sara Lyons and Amy Sackville, all of whom read and commented on an early manuscript, catching and querying errors where they could. Jane Graham-Maw, non-fiction supremo and agent-extraordinaire; thanks to her and all at GMC for savvy guidance, advice and all-round efficiency with a hearty good dose of kindness.

Many experts gave up their time to be interviewed, share their expertise and later to cast their eyes over what I had written, thanks to: Gary Ward, Dr Gianni Pes, Prof Graham Rook, Prof Tracy Kivell, Dr Irakli Loladze, Dr Matt Skinner and Prof Mike Adams – usual disclaimers apply and persisting errors are of course my own. Prof Wendy Parkins and Catherine Brereton also read and commented on sections. (Alas, there's still not very much about goats, Wendy.) David Surman lent the fine service of his pen for the illustrations.

This book was written in studies, offices, on trains, countless coffee shops, hotel rooms, sofas, pubs and libraries, even under a duvet or two, but kilos of gratitude go to Emma and Oliver Balch for lending me a room of one's own (and occasionally a shed, too) in which to start the book; my good friend Dr Nicola Ibba patiently put up with his visitor setting up camp beneath the air conditioning unit in 50 degree heat in Sardinia. Later on in the process Lynne Truss lent me a room and a shed, too – many thanks to all of you for making the writing of this a little easier and for the years of encouragement and friendship.

The people at Literature & Latte and at DEVONthink both work wonders on the other side of a computer screen and make writing this kind of book both easier and more pleasurable (thanks also go to Dr Christopher Mayo for introducing me to the latter).

Finally, there are numerous others who have keenly asked about my progress or provided help, support or reassurance: Erika, Adam, Ralph and Liberty Sinclair; Rebecca, Mike, Lloyd, Elliot and Scarlet Fairhurst; John, Lauren and Natasha Reid; (surrogate sister) Sandra Cryan, Julian Schutz, Sean and Martin Adkins; and also Huw Bevan, Ian Gouldstone, Alan Jenkins, Mehmet (scissorhands) Koraltan, Kate McAll at Pier Productions, Aylla MacPhaill, Kevin Mousley at the BBC, Florian Stadtler, Prof Scarlett Thomas, (the warrior himself) Thomas G Waites for teaching me how to read, David Flusfeder and Johanna Reid, my Mum, for doing all the things good mums do. Adam! What can I say about you, Adam? You, who watched practically every day of this book happen, saw it interfere with holidays and weekends, and listened with interest to every fact and speculation, and never tired of any of it. What have I done to deserve you? And finally, to Siân Prime, to whom this book is dedicated, who sets the standard by which all friendships should be measured.

PICTURE CREDITS

13 Metropolitan Museum of Art, online collection CC0; 32, 34 Author: Dbachmann, CC BY-SA 4.0, from Wikimedia Commons; 71 from "Ohalo II H2: A 19,000-year-old skeleton from a water-logged site at the Sea of Galilee, Israel", by I Hershovitz, M S Speirs, D Frayer *et al*, published in American Journal of Physical Anthropology © 2005 John Wiley and Sons; 75l Des Bartlett/Science Photo Library; 75r Sabena Jane Blackbird/ Alamy Stock Photo; 128, 162 & 163 from *On Deformities of the Spine*, by Edward W Duffin, John Churchill, London 1848; 141l DeAgostini/Getty Images; 141r Vybarr Cregan-Reid; 145 CC0, from Wikimedia Commons; 148 liliegraphie/123RF; 168 Science Museum, London/Wellcome Collection CC BY SA 4.0; 232, 233 from "Hidden shift of the ionome of plants exposed to elevated CO_2 depletes minerals at the base of human nutrition" by Irakli Loladze, eLife 2014; 3:e02245 DOI: 10.7554/ eLife.02245 CC SA © 2014, Loladze; 236, 275 l & r Vybarr Cregan-Reid; 277 from "Human hand-like use in *Australopithecus Africanus* – finger density", by Matthew M Skinner et al, 2015. Reprinted with permission from The American Association for the Advancement of Science